Praise for *Proofiness*

"Passionate . . . This is more than a math book; it's an eye-opening civics lesson." —*The New York Times Book Review*

"Spirited . . . Seife's book is an admirable salvo against quantitative bamboozlement by the media and the government. One hopes it will serve as a public inoculation against the malady it describes." —*The Boston Globe*

"Detailed and hard-hitting."

—*The Charlotte Observer*

"If Stephen Colbert had had time to write a math book, he surely would have written *Proofiness*."

—*The Dallas Morning News*

"Seife's coinages, humor, and curious tidbits keep readers engaged as the book gradually moves from a description of techniques to their practical application."

—*The Philadelphia Inquirer*

"Sprightly written, despite its sobering message."

—*Kirkus Reviews*

"A delightful and remarkably revealing book that should be required reading for . . . well, for everyone."

—*Booklist* (starred review)

PENGUIN BOOKS

PROOFINESS

Charles Seife is the author of four previous books, including *Sun in a Bottle* and *Zero*, which won the PEN/ Martha Albrand Award for first nonfiction book and was named a *New York Times* Notable Book. His work has appeared in such publications as the *New York Times*, *New Scientist*, *Scientific American*, the *Economist*, and *Wired*. He lives in New York City and is a professor of journalism at New York University.

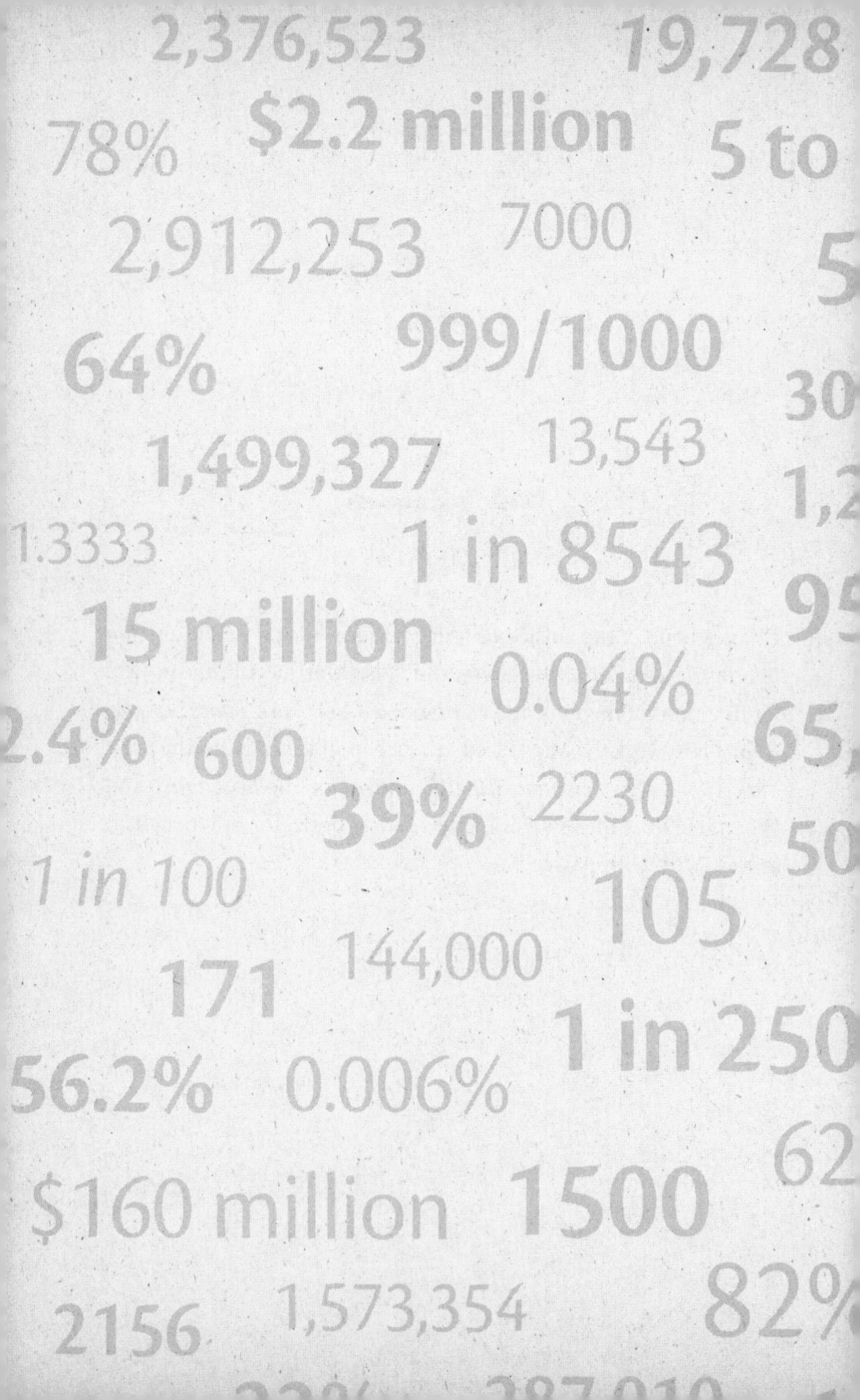

2,376,523
19,728
78%
$2.2 million
2,912,253
7000
64%
999/1000
1,499,327
13,543
1 in 8543
15 million
0.04%
600
39%
2230
1 in 100
105
171
144,000
56.2%
0.006%
$160 million
1500
2156
1,573,354

PROOFINESS

HOW YOU'RE BEING FOOLED BY THE NUMBERS

CHARLES SEIFE

PENGUIN BOOKS

PENGUIN BOOKS
Published by the Penguin Group
Penguin Group (USA) Inc., 375 Hudson Street, New York, New York 10014, U.S.A.
Penguin Group (Canada), 90 Eglinton Avenue East, Suite 700, Toronto,
Ontario, Canada M4P 2Y3 (a division of Pearson Penguin Canada Inc.)
Penguin Books Ltd, 80 Strand, London WC2R 0RL, England
Penguin Ireland, 25 St. Stephen's Green, Dublin 2, Ireland (a division of Penguin Books Ltd)
Penguin Books Australia Ltd, 250 Camberwell Road, Camberwell,
Victoria 3124, Australia (a division of Pearson Australia Group Pty Ltd)
Penguin Books India Pvt Ltd, 11 Community Centre,
Panchsheel Park, New Delhi – 110 017, India
Penguin Group (NZ), 67 Apollo Drive, Rosedale, Auckland 0632,
New Zealand (a division of Pearson New Zealand Ltd)
Penguin Books (South Africa) (Pty) Ltd, 24 Sturdee Avenue,
Rosebank, Johannesburg 2196, South Africa

Penguin Books Ltd, Registered Offices: 80 Strand, London WC2R 0RL, England

First published in the United States of America by Viking Penguin,
a member of Penguin Group (USA) Inc. 2010
Published in Penguin Books 2011

1 3 5 7 9 10 8 6 4 2

ILLUSTRATION CREDITS
Figure 1: Quaker Oats (redrawn for clarity)
Figures 5 and 6: *Journal of Neuropathology & Experimental Neurology*
(Figure 6 redrawn for clarity)
Figures 7 and 9: Nature Publishing Group
Figure 8: Dow Jones & Company, Inc.

THE LIBRARY OF CONGRESS HAS CATALOGED THE HARDCOVER EDITION AS FOLLOWS:
Seife, Charles.
Proofiness : the dark arts of mathematical deception / Charles Seife.
p. cm.
Includes bibliographical references and index.
ISBN 978-0-670-02216-8 (hc.)
ISBN 978-0-14-312007-0 (pbk.)
1. Mathematics—Miscellanea. 2. Pseudoscience. I. Title.
QA99.S45 2010
510—dc22 2010012127

Printed in the United States of America

Contents

PROOFINESS

INTRODUCTION

Proofiness

The American mind seems extremely vulnerable to the belief that any alleged knowledge which can be expressed in figures is in fact as final and exact as the figures in which it is expressed.

—Richard Hofstadter, *Anti-Intellectualism in American Life*

"In my opinion the State Department, which is one of the most important government departments, is thoroughly infested with communists."

This was not the sentence—delivered to a small gathering of West Virginia women—that catapulted the little-known Wisconsin senator into the public spotlight. It was the next one.

As he held aloft a sheaf of papers, a beetle-browed Joe McCarthy assured his place in the history books with his bold claim: "I have here in my hand a list of 205—a list of names that were made known to the Secretary of State as being members of the Communist party and who nevertheless are still working and shaping policy in the State Department."

That number—205—was a jolt of electricity that shocked Washington into action against communist infiltrators. Never mind that the number was a fabrication. It went up to 207 and then dropped again the following day, when McCarthy wrote to President Truman claiming that "we have been able to compile a list of 57 Communists in the State Department." A few days later, the number stabilized at 81 "security risks." McCarthy gave a lengthy speech in the Senate, giving some details about a large number of cases (fewer than 81, in fact), but without revealing enough information for others to check into his assertions.

It really didn't matter whether the list had 205 or 57 or 81 names. The very fact that McCarthy had attached a number to his accusations imbued them with an aura of truth. Would McCarthy make such specific claims if he didn't have evidence to back them up? Even though White House officials suspected that he was bluffing, the numbers made them doubt themselves.* The numbers gave McCarthy's accusations heft; they were too substantial, too specific to ignore. Congress was forced to hold hearings to attempt to salvage the reputation of the State Department—and the Truman administration.

McCarthy was, in fact, lying. He had no clue whether the State Department was harboring 205 communists or 57 or none at all; he was making wild guesses based upon information that he knew was worthless. Yet once he made the claim public and the Senate

* One presidential aide's notes reveal the confusion in the White House: "205 or 207? State Department can't find the names; we assume McCarthy doesn't have them either. 57? Said by McCarthy to be Communists. We don't know whether McCarthy has a list. He says he does. We don't know to whom he refers."

declared that it was going to hold hearings on the matter, he suddenly needed some names. So he approached newspaper magnate William Randolph Hearst, an ardent anticommunist, to help him compile a list. As Hearst recalled, "Joe never had any names. He came to us. 'What am I gonna do? You gotta help me.' So we gave him a few good reporters."

Even the assistance of half a dozen Hearst reporters and columnists couldn't give much substance to McCarthy's list. When the hearings began in March 1950, he couldn't produce the name of a single communist working for the State Department. It didn't make any difference. McCarthy's enumerated accusations had come at just the right time. China had just gone communist, and as the hearings finished, North Korea invaded the South. The United States was terrified of the rising tide of world communism, and McCarthy's bluff turned him, virtually overnight, into a symbol of resistance. An obscure back-bench junior senator had become one of the most famous and most divisive figures in politics. His line about 205 communists was one of the most effective political lies in American history.

The power of McCarthy's speech came from a number. Even though it was a fiction, that number lent credibility to the lies he was telling. It implied that the sheaf of papers he held in his hand was full of damning facts about specific State Department employees. The number 205 seemed to be powerful "proof" that McCarthy's accusations had to be taken seriously.

As McCarthy knew, numbers can be a powerful weapon. In skillful hands, phony data, bogus statistics, and bad mathematics can make the most fanciful idea, the most outrageous falsehood seem true. They can be used to bludgeon enemies, to destroy critics, and to squelch debate. Indeed, some people have become incredibly adept at using fake numbers to prove falsehoods. They have become masters of *proofiness*: the art of using bogus mathematical arguments to prove something that you know in your heart is true—even when it's not.

Our society is now awash in proofiness. Using a few powerful techniques, thousands of people are crafting mathematical falsehoods to get you to swallow untruths. Advertisers forge numbers to get you to buy their products. Politicians fiddle with data to try to get you to reelect them. Pundits and prophets use phony math to get you to believe predictions that never seem to pan out. Businessmen use bogus numerical arguments to steal your money. Pollsters, pretending to listen to what you have to say, use proofiness to tell you what they want you to believe.

Sometimes people use these techniques to try to convince you of frivolous and absurd things. Scientists and others have used proofiness to show that Olympic sprinters will one day break the sound barrier and that there's a mathematical formula that determines who has the perfect butt. There's no limit to how absurd proofiness can be.

At the same time, proofiness has extraordinarily serious consequences. It nullifies elections, crowning victors who are undeserving—both Republican and Democratic. Worse yet, it is used to fix the outcome of future elections; politicians and judges use wrongheaded mathematics to manipulate voting districts and undermine the census that dictates which Americans are represented in Congress. Proofiness is largely responsible for the near destruction of our economy—and for the great sucking sound of more than a trillion dollars vanishing from the treasury. Prosecutors and justices use proofiness to acquit the guilty and convict the innocent—and even to put people to death. In short, bad math is undermining our democracy.

The threat is coming from both the left and the right. Indeed, proofiness sometimes seems to be the only thing that Republicans and Democrats have in common. Yet it's possible to counteract it. Those who have learned to recognize proofiness can find it almost everywhere, ensnaring the public in a web of transparent falsehoods. To the wary, proofiness becomes a daily source of great amusement—and of blackest outrage.

Once you know the methods people use to turn numbers into falsehoods, they are powerless against you. When you learn to shovel proofiness out of the way, some of the most controversial topics become simple and straightforward. For example, the question of who actually won the 2000 presidential election becomes crystal clear. (The surprising answer is one that almost nobody would have been willing to accept: not Bush, not Gore, and almost none of the people who voted for either candidate.) Understand proofiness and you can uncover many truths that had been obscured by a haze of lies.

1

Phony Facts, Phony Figures

Facts are stubborn things; and whatever may be our wishes, our inclinations, or the dictates of our passion, they cannot alter the state of facts and evidence.

—John Adams

Facts are stupid things.

—Ronald Reagan

If you want to get people to believe something really, really stupid, just stick a number on it. Even the silliest absurdities seem plausible the moment that they're expressed in numerical terms.

Are blonds an endangered species? A few years ago, the media were all abuzz about a World Health Organization study proving that natural blonds would soon be a thing of the past. The BBC declared that people with blond hair "will become extinct by 2202." *Good Morning America* told its viewers that natural blonds will "vanish from the face of the earth within two hundred years" because the blond gene is "not as strong a gene as brunettes'." The story was winging its way around the globe until the WHO issued an unusual statement:

> WHO wishes to clarify that it has never conducted research on this subject. Nor, to the best of its knowledge, has WHO issued a report predicting that natural blondes are likely to be extinct by 2202. WHO has no knowledge of how these news reports originated but would like to stress that we have no opinion on the future existence of blondes.

It should have been obvious that the story was bogus, even before the WHO denial. One geneticist had even told the BBC as much. "Genes don't die out unless there is a disadvantage to having that gene," he said. "They don't disappear." But the BBC had suspended its faculties of disbelief. The reason, in part, was because of a phony number. The specificity, the seeming mathematical certainty of the prediction of when the last blond would be born, gave the story an aura of plausibility. It suckered journalists who should have known better.

No matter how idiotic, how unbelievable an idea is, numbers can give it credibility. "Fifty-eight percent of all the exercise done in America is broadcast on television," MSNBC host Deborah Norville declared in 2004, with a completely straight face. "For instance, of the 3.5 billion sit-ups done during 2003, two million, three hundred thousand [*sic*] of them were on exercise shows." Without once pausing to think, Norville swallowed the bogus statistics and regurgitated them for her audience; just a moment's reflection should have revealed that the story was nonsense.* The numbers had short-circuited Norville's brain, rendering her completely incapable of critical thought. It's typical. Numbers have that power over us, because in its purest form, a number is truth.

The cold and crystalline world of numbers gives us the rarest of all things: absolute certainty. Two plus two is always four. It was always so, long before our species walked the earth, and it will be so long after the end of civilization.

But there are numbers and there are numbers. Pure numbers are the domain of mathematicians—curious people who study numbers in the abstract, as Platonic ideals that reveal a higher truth. To a mathematician, a number is interesting in its own right. Not so for the rest of us.

For a nonmathematician, numbers are interesting only when they give us information about the world. A number only takes on any significance in everyday life when it tells us how many pounds we've gained since last month or how many dollars it will cost to buy a sandwich or how many weeks are left before our taxes are due or how much money is left in our IRAs. We don't care about the properties of the number five. Only when that number becomes attached to a unit—the "pounds" or "dollars" or "weeks" that signify what real-world property the number represents—does it become interesting to a nonmathematician.

A number without a unit is ethereal and abstract. With a unit, it acquires a meaning—but at the same time, it loses its purity. A number with a unit can no longer inhabit the Platonic realm of

---

* A few months later, perhaps unwilling to be outdone by his colleague, MSNBC host Keith Olbermann touted "a five-year study just concluded at Indiana University" which proved that "upon the birth of their first child, 100 percent of parents lose at least 12 IQ points, and the average loss is 20." These numbers, too, are fiction.

absolute truth; it becomes tainted with the uncertainties and imperfections of the real world. To mathematicians, numbers represent indisputable truths; to the rest of us, they come from inherently impure, imperfect measurements.

This uncertainty is unavoidable. Every unit represents an implied measurement. Inches, for example, represent an implied measurement of length; when someone says that a coffee table is eighteen inches wide, he's saying that if we were to take the table and measure it with a ruler, the table would have the same length as eighteen of the little hash marks we call inches. When someone says he weighs 180 pounds, he's saying that if you measured him with a bathroom scale, the number on the dial would read 180. Every number that has a real-world meaning is tied, at least implicitly, to a measurement of some kind. Liters are tied to a measurement of volume. Acres imply a measurement of area. Watts imply a measurement of power. A measurement of speed is expressed in miles per hour or in knots. A measurement of wealth is in dollars or euros or yuan. If someone says that he has five fingers, he's saying that if you count his digits—and counting objects is a measurement too—the answer will be five fingers.

It's universal; behind every real-world number, there's a measurement. And because measurements are inherently error-prone (they're performed by humans, after all, using instruments made by humans), they aren't perfectly reliable. Even the simple act of counting objects is prone to error, as we shall see. As a result, every measurement and every real-world number is a little bit fuzzy, a little bit uncertain. It is an imperfect reflection of reality. A number is always impure: it is an admixture of truth, error, and uncertainty.

Proofiness has power over us because we're blind to this impurity. Numbers, figures, and graphs all have an aura of perfection. They seem like absolute truth; they seem indisputable. But this is nothing but an illusion. The numbers of our everyday world—the numbers we care about—are flawed, and not just because measurements are imperfect. They can be changed and tinkered with, manipulated and spun and turned upside down. And because those lies are clad in the divine white garb of irrefutable fact, they are incredibly powerful. This is what makes proofiness so very dangerous.

It's true: all measurements are imperfect. However, some are more imperfect than others. As a result, not all numbers are equally fallible. Some numbers, those that are based upon extremely reliable and objective measurements, can come very close to absolute truth. Others—based on unreliable or subjective or nonsensical measurements—come close to absolute falsehood. It's not always obvious which are which.

Truthful numbers tend to come from good measurements. And a good measurement should be reproducible: repeat the measurement two or ten or five hundred times, you should get pretty much the same answer each time. A good measurement should also be objective. Even if different observers perform the measurement with different kinds of measuring devices, they should all agree about the outcome. A measurement of time or of length, for example, is objective and reproducible. If you hand stopwatches to a dozen people watching the same event—say, the Kentucky Derby—and ask them to time the race, they'll all come up with roughly the same answer (if they're competent). A whole stadium full of people, each using different stopwatches and clocks and time-measuring devices, would agree that the race took, say, roughly one minute and fifty-nine seconds to complete, give or take a few fractions of a second. Similarly, ask a dozen people to measure an object like a pencil, and it doesn't matter whether they use a ruler or a tape measure or a laser to gauge its length. When they complete their measurements, they'll all agree that the pencil is, say, four and a half inches long, give or take a fraction of an inch. The result of the measurement doesn't depend on who's doing the measuring or what kind of equipment's being used—the answer is always roughly the same. This is an essential property of a good measurement.

Bad measurements, on the other hand, deceive us into believing a falsehood—sometimes by design. And there are lots of bad measurements. Luckily, there are warning signs that tell you when a measurement is rotten.

One red flag is when a measurement attempts to gauge something that's ill-defined. For example, "intelligence" is a slippery concept—nobody can nail down precisely what it means—but that doesn't stop people from trying to measure it. There's an entire industry devoted to trying to pin numbers to people's brains; dozens and dozens of tests purport to measure intelligence, intellectual ability, or aptitude. (An applicant to Mensa, the high-IQ society, has his choice of some thirty-odd exams to prove his intellectual superiority.) Testing is just the tip of the multimillion-dollar iceberg. After measuring your intelligence, some companies sell you a set of exercises that help you improve your score on their tests, "proving" that you've become smarter. Dubious claims are everywhere: video games, DVDs, tapes, and books promise to make you more intelligent—for a price. Even the British Broadcasting Company tried to cash in on the intelligence-enhancement fad. In 2006, a BBC program promised that you can become "40 percent cleverer within seven days" by following diet advice and doing a few brain-teasers. Was there a sudden surge in the number of Britons understanding quantum physics? Unlikely. So long as researchers argue about what intelligence is, much less how to measure it, you can be assured that the "40 percent cleverer" claim is worthless. In fact, I can personally guarantee that you'll instantly be 63 percent smarter if you ignore all such statements.

Even if a phenomenon has a reasonable definition that everybody can agree about, it's not always easy to measure that phenomenon. Sometimes there's no settled-upon way to measure something reliably—there's no measuring device or other mechanism that allows different observers to get the same numbers when trying to quantify the phenomenon—which is another sign that the measurement is dubious. For example, it's hard to measure pain or happiness. There's no such thing as a painometer or a happyscope that can give you a direct and repeatable reading about what a subject is feeling.* In lieu of devices that can measure these experiences directly, researchers are forced to use crude and unreliable methods to try to get a handle on the degree of pain or happiness that a subject is feeling. They use questionnaires to gauge how much pain someone is in (circle the frowny face that corresponds with your level of pain) or how good someone feels (circle the number that represents how happy you are). Making matters even more difficult, pain and joy are subjective experiences. People feel them differently—some people are extremely tolerant to pain and some are very sensitive; some are emotional and regularly climb up

* This doesn't stop scientists from trying. To measure the effectiveness of painkillers in mice, some scientists use a calibrated hotplate; they measure pain by timing how long it takes for the mouse to jump or otherwise react to the hot surface.

towering peaks of bliss while others are more even-keeled. This means that even if a scientist could somehow devise an experiment where people would experience exactly the same amount of pain or joy, they would almost certainly give different answers on the questionnaire because their perceptions are different. A swift kick to the shins will elicit a super-duper frowny face from someone who has a low pain tolerance, while a more stoic person would barely deviate from a mild grimace. When rational people will come up with different answers to a question—how painful a blow to the head is, how beautiful a person in a photo is, how easy a book is to read, how good a movie is—the measurement can have some value, but the number is certainly far from the realm of absolute truth.

But it's not the farthest away. That honor goes to numbers that are tied to phony measurements—measurements that are fake or meaningless or even nonexistent. Numbers like these are everywhere, but product labels seem to be their favorite habitat. Just try to imagine what kind of measurement L'Oreal made to determine that its Extra Volume Collagen Mascara gives lashes "twelve times more impact." (Perhaps they had someone blink and listened to how much noise her eyelashes made when they clunked together.) How much diligence do you think Vaseline put into its research that allowed it to conclude its new moisturizer "delivers 70 percent more moisture in every drop." (Presumably it would deliver less moisture than water, which is 100 percent moisture, after all.) No matter how ridiculous an idea, putting it into numerical form makes it sound respectable, even if the implied measurement is transparently absurd. This is why paranormal researchers feel compelled to claim, without giggling, that 29 percent of Christian saints had exhibited psychic powers.

Making up scientific-sounding measurements is a grand old

tradition; cigarette companies used to excel at the practice, the better to fill their ads with a thick haze of nonsense. "From first puff to last, Chesterfield gives you a smoke *measurably* smoother . . . cooler . . . best for you!" read one advertisement from 1955. You can't measure the smoothness and coolness of a cigarette any more than you can measure the impact of an eyelash. Even if people *tried* to quantify impact or smoothness or coolness, the results would be worthless. These are phony measurements. They're like actors dressed up in lab coats—they appear to be scientific, but they're fake through and through. As a result, the numbers associated with these measurements are utterly devoid of meaning. They are fabricated statistics: *Potemkin numbers.*

According to legend, Prince Grigory Potemkin didn't want the empress of Russia to know that a region in the Crimea was a barren wasteland. Potemkin felt he had to convince the empress that the area was thriving and full of life, so he constructed elaborate façades along her route—crudely painted wooden frameworks meant to look like villages and towns from afar. Even though these "Potemkin villages" were completely empty—a closer inspection would reveal them to be mere imitations of villages rather than real ones—they were good enough to fool the empress, who breezed by them without alighting from her carriage.

Potemkin numbers are the mathematical equivalent of Potemkin villages. They're numerical façades that look like real data. Meaningful real-world numbers are tied to a reasonably solid measurement of some sort, at least implicitly. Potemkin numbers aren't meaningful because either they are born out of a nonsensical measurement or they're not tied to a genuine measurement at all, springing forth fully formed from a fabricator's head.

For example, on October 16, 1995, Louis Farrakhan, leader of the Nation of Islam, held an enormous rally: the "Million Man March." Of course, the gathering was named long before anyone knew whether a million men would actually attend the rally. When buses and trains and planes filled with men began converging on the National Mall in Washington, D.C., it was a huge event, but did it really live up to its name? Quite naturally, Farrakhan said that it did; his unofficial count of the crowd topped one million people. However, Farrakhan's count was a Potemkin number. It was a foregone conclusion that the organizers of the Million Man March would declare that a million men were in attendance, regardless of how many actually attended—anything else would be embarrassing. The Park Service, which was in charge of giving an official estimate of the crowd size, was already feeling the pressure to inflate the numbers. "If we say [the crowd] was 250,000, we'll be told it was a half-million," a U.S. Park Police officer told the *Washington Post* shortly before the rally. "If we say it was a half-million, we'll be told it was a million. Anything short of a million, and you can probably bet we'll take some heat for it." Nevertheless, the Park Service dutifully peered at aerial photos and counted heads in an attempt to size up the crowd. As predicted, when the official tally came in—400,000 people, give or take 20 percent or so—a furious Farrakhan threatened to sue.* As a result, the Park Service stopped estimating crowd sizes for more than a decade, giving rally organizers free rein to make up Potemkin-number crowd estimates without fear of contradiction from a reliable source.

Whenever there's a big public event, someone has a vested inter-

* Though the Park Service's estimate was probably the best out there, some scholarly estimates went as high as roughly 900,000 people—still short of a million, but more than double the official estimates. This just goes to show that even the seemingly simple act of counting objects can be difficult—and politically sensitive.

est in making up a number that makes the crowd look huge. Pretty much every year, the organizers of Pasadena's Rose Parade announce (presumably for the ears of sponsors) that they estimate a crowd of one million spectators; the real number is probably half that. Estimates of the crowd at Barack Obama's inauguration topped five million at one point—probably between double and triple the number of people actually in attendance. The right is just as guilty as the left of fudging crowd numbers. In September 2009, a number of right-wing commentators claimed that ABC News had estimated that antiadministration "tea party" protests in D.C. had drawn more than a million people; in fact, ABC had pegged the crowd at between 60,000 and 70,000 attendees. Two months later, Republican representative Michele Bachmann bragged on the conservative *Sean Hannity Show* that a protest she organized had attracted between 20,000 and 45,000 angry citizens—and a video montage of swirling crowds seemed to support her assertions. However, the estimate was way off—the *Washington Post* estimated that the crowd was about 10,000. Worse yet, some of *The Sean Hannity Show*'s footage was recycled from the much larger September protest, making the crowd look much more substantial than it actually was.*

Creators of Potemkin numbers care little about whether their numbers are grounded in any sort of reality. From afar, however, they seem convincing, and a Potemkin number can be a powerful tool to prop up a sagging argument or to bludgeon a stubborn opponent. Even the flimsiest of them can do tremendous damage. Joe McCarthy's famous claim to know of 205 communists in the State

---

* After the deception was exposed by comedian Jon Stewart on *The Daily Show*, Sean Hannity promptly admitted that he had made an "error": "We screwed up. It was an inadvertent mistake, but it was a mistake nonetheless." Then, smirking, he thanked Stewart for watching the show.

Department, for example, was transparently false, yet it made him a national figure.

Using Potemkin numbers is the most overt form of proofiness. It takes the least skill to pull off. After all, it's incredibly easy to fabricate a number that suits whatever argument you are trying to make. (This is why Potemkin numbers are so common: 78 percent of all statistics are made up on the spot.)* However, the power of Potemkin numbers is limited. It begins to evaporate as soon as someone is thoughtful enough to examine the numbers more carefully. There are other more subtle—and dangerous—forms of proofiness to watch out for.

There's an anecdote about an aging guide at a natural history museum. Every day, the guide gives tours of the exhibits, without fail ending with the most spectacular sight in the museum. It's a skeleton of a fearsome dinosaur—a tyrannosaurus rex—that towers high over the wide-eyed tour group. One day, a teenager gestures at the skeleton and asks the guide, "How old is it?"

"Sixty-five million and thirty-eight years old," the guide responds proudly.

"How could you possibly know that?" the teenager shoots back.

"Simple! On the very first day that I started working at the museum, I asked a scientist the very same question. He told me that the skeleton was sixty-five million years old. That was thirty-eight years ago."

It's not a hilarious anecdote, but it illustrates an important point. The number the museum guide gives—sixty-five million and thirty-

* And 36 percent of readers will actually believe that statistic.

meter or so. No matter how careful you are, no matter how many times you measure and remeasure, no matter how carefully you squint through a magnifying glass when you read the hash marks, there's no way you can use a ruler to get an answer much better than "roughly 150.1 millimeters" for the length of a pencil. That's as precise as you can get with a ruler.

If you listen carefully enough, numbers tell you that they're only approximations. They reveal their limitations—better yet, they tell you how far to trust them. This information is encoded in the way we talk about numbers; it's already second nature to you, even though you might not recognize it. When someone declares that a pencil is 150.112 millimeters long, you automatically assume that the measurement is extremely precise. On the other hand, if he says it is 150 millimeters long, you would assume that the measurement is much rougher. Nice round numbers are sending a subliminal signal that their associated measurements have large errors—the numbers are announcing that you can't trust them very far because they're crude approximations. Long, ultra-specific numbers send exactly the opposite message: that they come from measurements that are more trustworthy and closer to absolute truth. All real-world numbers behave like this. When someone tells you that his car cost $15,000, you automatically assume that there's quite a bit of slop in the figure—the real cost was somewhere in the ballpark of fifteen grand, give or take a few hundred dollars. Yet if someone says that his car cost $15,232, you then assume that this was the precise amount he paid, give or take a few pennies. Similarly, if someone tells you that he's eighteen years old, you expect that he's somewhere between eighteen and nineteen years of age. If he says that he's eighteen years, two months, and three days old, you know that his answer is good to within a few hours—and that he's probably a bit obsessive. The roundness of a number gives you clues about how precise the number is, and how seriously you can take it.

This is the key to the dinosaur anecdote. When a scientist says that a dinosaur skeleton is sixty-five million years old, it's a signal that the number is a fairly rough approximation; the measurement error is on the order of tens or hundreds of thousands of years.* In reality, the skeleton might be 64,963,211 years old; perhaps it's 65,031,844 years old. However, the paleontologist's measurements weren't precise enough to reveal that truth. When he said that the skeleton was sixty-five million years old, he was admitting that his measurement had large errors—it was sixty-five million years old, give or take tens or hundreds of thousands or even millions of years.

The museum guide screwed up when he took the sixty-five-million-year figure too literally. He ignored the errors inherent to the measurements of the dinosaur's age—the errors signaled by the roughness of the figure—and instead assumed that the skeleton was *exactly* sixty-five million years old when he began work at the museum. Only then would his hyper-precise figure of 65,000,038 years make sense. But since the errors in measurement absolutely dwarf the time he spent working at the museum, his figure of 65,000,038 years is ridiculous. The skeleton is still sixty-five million years old—as it will be a hundred or a thousand years in the future. The

* A great deal of scientific effort goes into increasing the precision of measurements to get closer to the truth by adding an extra decimal place or two. For example, in the past two decades, scientists' estimate of the age of the universe went from "about 15 billion years" to "14 billion years" to "13.7 billion years." This seemingly subtle change represents an extraordinary—almost revolutionary—advance in our knowledge about the cosmos.

guide erred because he trusted the measurement beyond the point where it should be trusted. He committed an act of *disestimation*.

Disestimation is the act of taking a number too literally, understating or ignoring the uncertainties that surround it. Disestimation imbues a number with more precision than it deserves, dressing a measurement up as absolute fact instead of presenting it as the error-prone estimate that it really is. It's a subtle form of proofiness: it makes a number look more truthful than it actually is—and the result can be as silly and meaningless as the museum guide's 65,000,038-year-old dinosaur.

Every few years, public officials and the news media perform a ritual form of disestimation when a population clock reaches a big milestone. Population experts at the Census Bureau and around the world are constantly estimating the populations of each nation. Their estimates are pretty good, predicting when, say, the world's population reaches six billion—they might even be able to guess when the six billionth person is born to within a few hours. That's about as good as any possible measurement of population can get. Populations constantly fluctuate, with people dying and being born at irregular intervals, often far from the eyes of people who count such things, so it's impossible to know at any given moment the true number of people alive on earth. Nevertheless, on October 13, 1999, as flashbulbs popped around him, UN secretary-general Kofi Annan held a young Bosnian boy, welcoming him into the world as the six billionth person on earth. (The UN insisted that Annan's presence in Sarajevo was a complete coincidence. It was just a lucky break that the six billionth person was born in the city where Annan happened to be visiting.)

There's no way that anyone could pinpoint which baby became the six billionth person living on earth. The uncertainties in measurement are simply too huge. You wouldn't know, probably to within several thousand, whether a baby is number 6,000,000,000 or 5,999,998,346 or 6,000,001,954. Only by disestimating, by ignoring the uncertainty in population numbers, could anyone claim to know for certain who was the six billionth living person. Yet at every population milestone, world officials and the news media go through the same bizarre pantomime. In 2006, the *Chicago Sun-Times* declared a local baby—Alyzandra Ruiz—to be the 300 millionth resident of the United States. (They cleverly jumped the gun on everybody, making the arbitrary call almost an hour before the official Census Bureau population estimate reached the 300 million mark.) And when the world population reaches seven billion, probably in early 2012, officials will declare some lucky baby to be the seven billionth living person, completely indifferent to the fact that it's a lie.

Disestimates have much more staying power than Potemkin numbers. While a Potemkin number is purely fanciful and intended to deceive, a disestimate has its origin in a real, meaningful, good-faith measurement—the problem is that we don't take the resulting number with a big enough grain of salt. It's a rather subtle problem. As a result, disestimates can be difficult to spot. And they don't wither under scrutiny like Potemkin numbers do. Once a disestimate is believed by the public, it can be devilishly hard to debunk. As an example, ask yourself: what body temperature is normal? If you live in the United States—one of the few countries left that still use the antiquated Fahrenheit scale—your answer almost certainly is 98.6 degrees Fahrenheit. If you have an (analog) medical thermometer in your medicine cabinet, it probably has a little arrow pointing to 98.6°F. When you see the little line of liquid creep beyond that arrow, you probably conclude that you've got a fever. What you might not know, though, is that 98.6°F is a disestimate.

The idea that normal body temperature is 98.6°F comes from

research done in the late 1860s by the German physician Carl Wunderlich. Even though this number seems very precise and official, that precision is an illusion. There are quite a few reasons not to take the 98.6°F number literally.

Wunderlich may have been faking his data. He made the (rather dubious) claim to have measured a million body temperatures with unlikely precision. After taking those temperatures, he came to the conclusion that "normal" temperature was 37 degrees Celsius—a nice round number in the temperature scale used by most of the world. Converting the nice round 37-degree Celsius number into Fahrenheit yields 98.6 degrees F, automatically making the number seem more precise than it actually is. Furthermore, Wunderlich took the body temperatures in his subjects' armpits, so even if his measurements were valid and precise, his definition of "normal" wouldn't apply to measurements of body temperature taken from the mouth or other orifices, as these have slightly different temperatures. Body temperature isn't uniform—the answer you get depends on where you make the measurement. This is a huge source of error that most people don't take into account. Neither do they seem to compensate for the fact that body temperatures can change dramatically throughout the day, and that "normal" is very different from person to person. There is no hard-and-fast definition of "normal," much less one that's precise to within a tenth of a degree as the 98.6°F number seems to be. Yet it's a fiction that we still cling to—even medical dictionaries sometimes define a fever as a body temperature above 98.6°F. We all imbue the highly precise number with tremendous importance, even though in truth the definition of "normal" temperature is imprecise, fuzzy, and somewhat arbitrary. It's a disestimation, yet one that's persisted for a century and a half.

Because of disestimation's subtlety and longevity, it can be a particularly nasty form of proofiness. Even though disestimates aren't complete nonsense in the way that Potemkin numbers are, they mix a whole lot of fiction in with their fact. Failure to recognize the inherent limitations of a measurement can be extremely dangerous, because it can potentially create an authentic-sounding number that is in fact far removed from the realm of truth.

There are many roads that lead to proofiness. Potemkin numbers create meaningless statistics. Disestimation distorts numbers, turning them into falsehoods by ignoring their inherent limitations. A third method, *fruit-packing*, is slightly different. In fruit-packing, it's not the individual numbers that are false; it's the presentation of the data that creates the proofiness.

Supermarkets select their fruit and arrange it just so and package it so that even mediocre produce looks delectable. Similarly, numerical fruit packers select data and arrange them and dress them up so that they look unassailable, even when they're questionable. The most skilled fruit packers can make numbers, even solid ones, lie by placing them in the wrong context. It's a surprisingly effective technique.

A particularly powerful weapon in the fruit packer's arsenal is what's known as *cherry-picking*. Cherry-picking is the careful selection of data, choosing those that support the argument you wish to make while underplaying or ignoring data that undermine it.

Since real-world numbers are fuzzy, answers to numerical questions aren't always clear-cut. Measuring the same thing in different ways can give different answers; some of the numbers will be too high, some will be too low, and, with luck, others will be reasonably close to the right answer. The best way to figure out where the truth lies is to look at all of the data together, figuring out the advantages and disadvantages of each kind of measurement so that you get as close to the truth as possible. A cherry picker, on the other hand, selects the data that support his argument and presents only them, willfully excluding numbers that are less supportive, even if those numbers may be closer to the truth. Cherry-picking is lying by exclusion to make an argument seem more compelling. And it's extremely common, especially in the political world.* Every politician is guilty of it, at least to some extent.

Al Gore is guilty of cherry-picking in his film *An Inconvenient Truth*. At the heart of the 2006 movie is a breathtaking and disturbing sequence where he shows computer simulations of what global warming will do to the surface of the earth. In a series of maps, he shows the world's coastlines disappearing under the rising oceans. Much of Florida and Louisiana will be submerged, and most of Bangladesh will sink beneath the waves. The animations are stunning, leaving viewers with little doubt that global warming will dramatically reshape our planet. However, those animations are based upon a cherry-picked number: Gore's pictures assume that melting ice will drive the sea level up by twenty feet.

Lots of scientists have tried to model the effects of global warming, and most have come to a very different conclusion. They tend to agree that global warming is real, that human activities are responsible for a sizable portion of that warming, and that the sea

---

* It's also very common in the scientific world, thanks to a phenomenon known as "publication bias." Peer-reviewed journals cherry-pick the most exciting papers, selecting them for publication. This means that papers with spectacular results are published in high-profile journals while less sexy ones (including negative results) are relegated to lesser journals or aren't published at all. Publication bias distorts science, making new drugs, for example, seem more effective than they actually are.

level will indeed rise over the next century.* There's an outside chance that the sea level will rise by twenty feet or more if a very-worst-case scenario occurs (such as the near-complete melting of the ice sheets in Greenland or West Antarctica). However, most serious estimates project a sea level rise much lower than what Gore used. Some climatologists say the oceans will rise two feet or so in the next century; some go as high as four feet—these are the best scientific guesses right now. Yet Gore ignores these more modest estimates and picks the most extreme model of sea level rise—the twenty-footer—so he can flash his dire graphics on the screen. It wowed the audiences, but it was cherry-picking.

George W. Bush is just as guilty of cherry-picking as his erstwhile opponent. Like every president, he put the best possible spin on all of his pet projects. "No Child Left Behind," for example, was the name for a shift in educational strategy early in his administration; the act, signed by Bush in 2002, offered money to states in return for mandatory testing and other concessions. It was a controversial move. Several years later, in his State of the Union address, Bush declared that No Child Left Behind was working wonders in America's schools. "Five years ago, we rose above partisan differences to pass the No Child Left Behind Act . . . and because we acted, students are performing better in reading and math." This statement was a rare instance of a double cherry-pick.

First, when Bush flatly declared that students are doing better in math and reading, he had to do a bit of cherry-picking. His data came from the Department of Education, which periodically spon-

* Despite my singling out Al Gore for cherry-picking, there are unambiguous data that show that global warming is occurring. It's just that sea levels aren't going to rise twenty feet anytime soon.

sors a national set of assessment tests to determine how well the nation's students are doing in various subjects. The data show that fourth- and eighth-grade students' reading and math scores have in fact improved since No Child Left Behind started. But there are other data that he ignored. Twelfth-grade students' reading scores declined over the same period. And though it's a little more complicated (the test changed form, making the trend harder to figure out), twelfth-grade math scores also seem to have declined slightly. So saying that students are performing better in reading and math is only true if you ignore the twelfth-grade results that say otherwise. Accentuate the positive; eliminate the negative. Cherry-pick number one.

Second, pretending that the improvement in math and reading scores is due to the No Child Left Behind Act requires some cherry-picking too. If you look at the scores carefully, you see that fourth- and eighth-grade math scores have been improving at roughly the same rate since the 1990s, long before the act was passed. Similarly, fourth- and eighth-grade reading scores have been improving at roughly the same (very modest) rate in the same time period. By ignoring data from before 2002, Bush was able to pretend that No Child Left Behind was responsible for the improving scores, even when it's clear that the trend is essentially unchanged over the years. No Child Left Behind only seems responsible for the improved scores if you fail to present earlier data that put the scores in the proper context. Cherry-pick number two.* Voilà. Bush can declare No Child Left Behind a success—even if it isn't.

---

* There's a third kind of cherry-pick here, in fact. Even if you accept that math and reading scores are important, schools teach a lot more subjects: writing, science, history, and more. Data from these disciplines show either mild improvement or, in some cases, decline, particularly in upper grades. Concentrating on reading and (particularly) math as indicators of a school's improvement is only looking at part of the picture.

Education statistics are a hotbed of fruit-packing. It's really hard to improve the school system; it requires lots of money and effort and time to make a change in a huge bureaucracy with such enormous inertia. Even more disturbing for a politician, it takes years before an administration can reap the benefits of making good educational policy; you might be out of office long before citizens realize that you've improved the school system. Fruit-packing provides a shortcut—it allows a politician to reap the benefits of improving the schools without ever having to do the hard work of changing policy.

In New York, scores on the state's reading and math tests have risen sharply since 2005. Any politician who has anything to do with education policy in the state basks in the glow of the rising scores each year. In 2008, for example, New York City mayor Michael Bloomberg declared that the "dramatic upward trend" in state test scores showed that the city's public schools were "in a different league" than when he took office. However, soon after the state tests were administered in 2005, teachers told reporters that the test was much easier than the year before. "What a difference from the 2004 test," a principal of a Bronx school told the *New York Times*. "I was so happy for the kids—they felt good after they took the 2005 test." Scores on the state tests climbed year after year, rising dramatically and improbably. On national tests, though, New York didn't seem like such a success story. In New York City, for example, scores on the national tests stayed more or less unchanged. It's pretty clear that New York State had been tinkering with the difficulty of the test. By making the test easier year after year, it artificially makes students' test scores rise. It seems as if the children are performing better on the tests, but in fact the rise in scores is meaningless; the 2004 test score doesn't mean the same thing as the 2006 test score, which doesn't mean the same thing as the 2008 test score. By pretending that these tests are equivalent, New York is engaging in another form of fruit-packing: *comparing apples to oranges.*

This particular trick is a game of units. As mentioned earlier in this chapter, every real-world number has a unit attached to it—a little tag like "feet" or "seconds" or "kilograms" or "dollars" that tells you what kind of measurement the number is tied to. When you compare two numbers to see which one's bigger than the other, it's important to ensure that the units are the same, otherwise the comparison is meaningless. Sometimes this is obvious (which is longer, 55 seconds or 6.3 feet?), but sometimes it's a little tricky to tell that the units aren't quite the same. Which is better: a 50 percent score on test A or a 70 percent score on test B? It's a meaningless comparison unless you have some way of converting a test A score into a test B score and vice versa.

There's no value to making a direct comparison of test scores from year to year unless the test makers ensure that the value of a given score always stays the same, yet this is precisely what New York State did, exploiting this apples-and-oranges problem to make it look as though their educational system were improving. The effect is very similar to cherry-picking; when New York compared apples to oranges, they distorted the meaning of numbers, making the statistics appear to support an argument that they don't.

Apple-orange comparisons can be really tough to spot, because units can be fluid creatures. Some of them change their meaning over time. The dollar, for example, is the unit that Americans use to measure money. But as a unit of wealth, it is always changing. Flip through an old magazine and look at the ads. The changing value of the dollar will hit you on the head. A December 1970 copy of

Esquire that I happen to have in my office shows that a low-end two-door car cost $1,899. Name-brand carry-on luggage would set you back $17. A pair of mass-produced men's shoes is worth $19. Right now (2010), the equivalent low-end two-door car costs about $12,000. The name-brand carry-on luggage will set you back $130. The pair of mass-produced men's shoes is worth roughly $100. Even though all of these numbers seem to have the same unit symbol—$—next to them, a dollar in 1970 is a very different unit than a dollar in 2010. In 1970, a dollar bought more than 5 percent of a pair of shoes. Now it buys less than 1 percent. Back then, "$" had a lot more value than it does today. If you look at the prices in that 1970s magazine carefully, you should quickly come to the conclusion that a 1970 dollar had somewhere between five and seven times more purchasing power than the 2010 dollar does. (Purchasing a car costs 6.3 times as many 2010 dollars as it did 1970 dollars, for example.) A 1970 dollar is a very different unit from a 2010 dollar, just as surely as a gallon is different from a quart.

As you probably know, this change is caused by inflation. Goods and services get a touch—typically around 3 percent—more expensive every year. Year after year, it takes more dollars to buy a gallon of paint, and it also takes more dollars to hire someone to paint your house, more dollars to take the bus to get to work, and more dollars to do just about everything you need money to do. Thus the almighty dollar gets less and less valuable as the years tick by. This makes it a little bit difficult to compare spending over time; you've got to adjust for inflation, taking into account the changing value of the dollar. However, sometimes people conveniently "forget" to make that conversion to make their arguments seem stronger. In November 2005, a group of House Democrats (the "Blue Dog Coalition") attacked George W. Bush with this little tidbit:

> Throughout the first 224 years (1776–2000) of our nation's history, 42 U.S. presidents borrowed a combined $1.01 trillion from foreign governments and financial institutions according to the U.S. Treasury Department. In the past four years alone (2001–2005), the Bush Administration has borrowed a staggering $1.05 trillion.

These figures are all true, and, yes, Bush was quite the deficit spender. However, this comparison is utterly meaningless because it's comparing apples to oranges—2005 and 2001 dollars are very different from 1776 and 1803 and 1926 dollars. The Louisiana Purchase cost $15 million; Alaska was a bargain at roughly half that. Back then, those were enormous expenditures that only wealthy states could afford. Nowadays the same sum is the cost of a fancy Manhattan penthouse. Past presidents who drove up the nation's debt did it in much smaller dollar figures simply because those dollar figures represented a lot more purchasing power than they do today.

Comparing apples and oranges can be quite powerful; used skillfully, it can make the false seem true and the true seem false. For example, in 2005, the director of the National Science Foundation gently bragged about the agency's budget request for the following year. The NSF, he said, would get "$132 million, or 2.4 percent, more than in [fiscal year] 2005. This modest increase allows us to assume new responsibilities, meet our ongoing commitments, and employ more staff. . . ." At first glance, a $132 million increase seems like something to celebrate. But the number was a fruit packer's fantasy. It came from comparing 2006 dollars to 2005 dollars without taking inflation into account. The 2006 dollar was worth less than the 2005 dollar, so the "increased" budget in 2006 would in fact be worth less than the budget in 2005. If you crunch the numbers properly, that $132 million increase was a disaster; in fact, the agency would be *losing* about $30 million (in 2006 dollars). There wouldn't be any extra money for new responsibilities or new staff. The NSF director was lying, turning a defeat into a victory.

By comparing apples with oranges, a skilled official can make a decrease look like an increase, up look like down, and black look like white. It's Orwellian—comparing apples to oranges can make fiction of fact and fact of fiction.*

Yet another variety of fruit-packing, *apple-polishing*, is used to put the finishing touches on data, manipulating them so they appear more favorable than they actually are. Just as greengrocers employ subtle artifices to make their produce look fresher and tastier than it actually is—waxing and polishing apples to make them look fresher, gassing tomatoes to make them turn red, piling cantaloupes so that their blemishes are hidden—mathematical fruit packers tweak their data, subtly polishing the numbers to make them look as appealing as possible.

There are endless ways to polish mathematical apples; it would be impossible to describe them all, especially since inventive fruit packers are inventing new ones all the time. But there are a few common tricks worth mentioning.

Graphs—visual depictions of data—are particularly vulnerable to apple-polishing. A fruit packer can choose to display data in an

* Budgets are always subject to proofiness. The people making them have a vested interest in making expenditures seem tiny; those criticizing them are trying to make those same numbers seem large. Though the details are too wonky for this book, it's worth mentioning that in the United States, the government keeps certain expenditures (like Social Security spending) off the official budget, counting them separately. Because of this, politicians can cherry-pick, ignoring off-budget expenditures when it suits them.

endless variety of ways, fiddling with the look so that the graph makes the data look more impressive than they actually are.

Take the case of Quaker Oats. It's a bland and relatively unappetizing product—not easy to come up with an ad campaign for. Yet people will eat anything that they think will improve their health, so ad executives launched a blitz to make the barely digestible oat fiber appear to be a medicinal vacuum cleaner, sucking the cholesterol right out of your bloodstream. They emphasized the point with a graph:

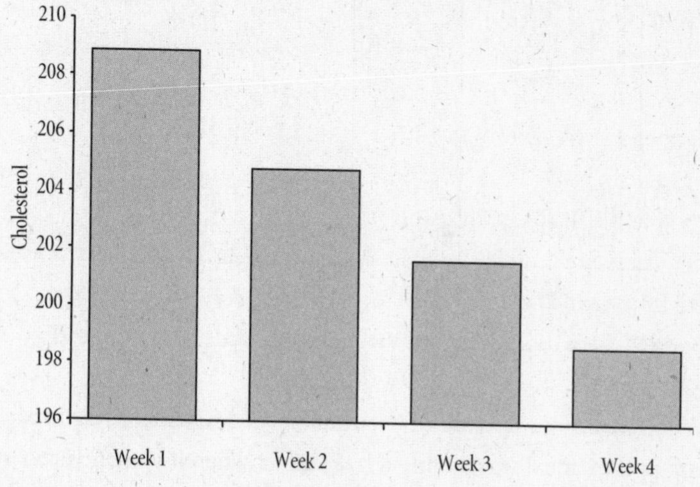

Figure 1. A deceptive graph about Quaker Oats.

The message was clear: eat Quaker Oats, and within a few weeks your cholesterol levels will drop dramatically. However, if you look carefully, you will discover that this graph is deceptive. We normally assume that the line at the bottom of the chart represents zero cholesterol—the little oat-fiber machines have gobbled up every single dollop of cholesterol in your blood. But if you examine the vertical axis of the chart, you see that the bottom isn't zero, but 196. This makes the data seem much more dramatic than they actually are, as you can easily see if you look at a more honest graph:

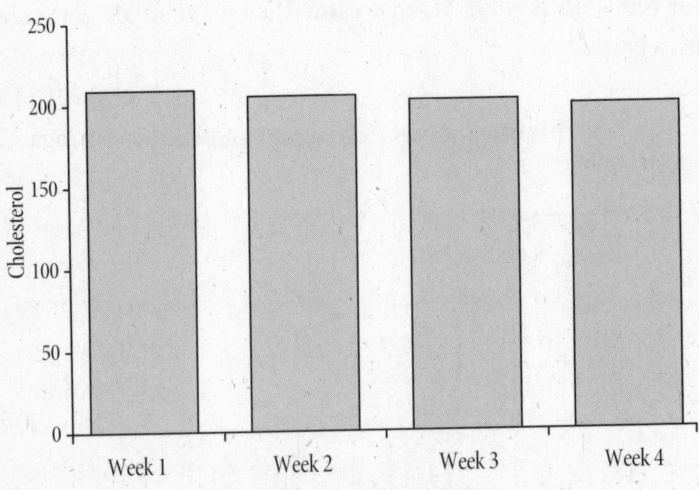

Figure 2. A less deceptive graph about Quaker Oats.

By tweaking the scale of the chart just so, Quaker Oats made it look as if oatmeal was having a huge effect when it wasn't. (After receiving complaints, Quaker withdrew the chart.) Of the many ways to manipulate data in graphs, this is probably the most common.

Another form of apple-polishing exploits the term "average" to make numbers seem smaller or larger than they really are. Most people think that "average" means "typical"—that if, say, the average salary at a company is $100,000, then each employee earns $100,000, more or less. In fact, that's often not the case.

The average of a set of numbers—more precisely, the *mean*—has a precise mathematical meaning: you add everything together and then divide by the number of data points that you added together. For example, if you had a company of ten people, each of whom earned roughly $100,000, you add those ten salaries together ($100,000 + $101,000 + $98,500 + $99,700 + $103,200 + $100,300 + $99,000 + $96,800 + $100,000 + $101,500 = $1,000,000) and then divide by the number of salaries ($1,000,000 ÷ 10 = $100,000). In this case, the average, $100,000, does in fact represent a typical salary. However, consider a company where the CEO earns $999,991 per year, and there are nine interns who each earn $1. The mean, again, is the sum of those salaries ($999,991 + $1 + $1 + $1 + $1 + $1 + $1 + $1 + $1 + $1 = $1,000,000) divided by the number of salaries ($1,000,000 ÷ 10 = $100,000). So here too the "average" salary is $100,000. However, $100,000 is not a "typical" salary in any meaningful way. If you were to pick a person at random from the company, you'd probably find that he earns a measly $1. So in this case, it's deceptive to pretend that "average" is "typical."* If the CEO were to recruit new employees by highlighting the company's average salary of $100,000, he would be apple-polishing. The new hire would be shocked when he gets his first paycheck.

Whenever a politician announces a tax cut, it's almost guaranteed that he'll pull the exact same stunt to make the tax breaks look larger than they actually are. He'll give a speech that talks about the "average" refund—the mean tax break—and make his constituents extremely happy. However, the "average" is usually far from typical.

---

* In cases like this, it's often better to use a construct known as the *median* to figure out what a typical salary should be. To calculate the median, you line the numbers up from lowest to highest and pick the one in the middle. Here, the median salary would be $1, clearly a better representation of "typical" than the mean would be.

Most people will be disappointed when they receive their refund checks in the mail. For example, George W. Bush made tax cuts a central thrust of his administration and always polished apples when describing them. A typical incident occurred at the end of his first term when he said, "The tax relief we passed, 11 million taxpayers this year will save $1,086 off their taxes." (The White House quickly corrected the figures; Bush meant to say 111 million taxpayers and an average of $1,586 in savings.) As it happens, though, both figures were deceptive. The typical family didn't see anywhere near $1,586 or $1,086 in tax breaks; most received less than $650. The reason was the same as the greedy-CEO example: a relatively small number of people received very large refunds, making the "average" very atypical. The rosy numbers coming out of the White House were technically true, but they were functionally lies—they were apple-polished to make them look much larger than they should.

Apple-polishing, cherry-picking, comparing apples to oranges—all the tricks of the fruit packer—present numbers in a misleading manner, distorting them to the point of falsehood. Potemkin numbers dress up nonsense in the guise of meaningful data. Disestimates stretch numbers beyond their breaking points, turning even valid measurements into lies. All of these techniques are forms of proofiness; all of them allow an unscrupulous person to make falsehoods look like numerical fact. And because we humans tend to think of numbers as representing absolute truth, we are hardwired to accept them without question.

Proofiness has such a hold over us because our minds are primed to accept mathematical falsehoods. Because of the way our brains work, certain kinds of numbers make them malfunction. As a result, we humans believe some absurd and embarrassing lies.

2

Rorschach's Demon

How easy it is to work over an undigested mass of data and emerge with a pattern, which at first glance, is so intricately put together that it is difficult to believe it is nothing more than the product of a man's brain.

—Martin Gardner, *Fads and Fallacies in the Name of Science*

Put down that credit card! Every time you make a purchase, you're destroying your health and tottering one step closer to the grave. "Researchers Link Bad Debt to Bad Health," blared the *New York Times* in 2000. Researchers at Ohio State had found that the deeper someone's debt, the worse his health was—leading them to conclude that credit card debt "could be bad for physical health" as well as financial health. "This is part of the dark side of the economic boom," said one of the scientists who studied the matter.

The newspapers are trying to convince us of a lot of silly things. Wearing red makes you perform better in sporting events. Smoking marijuana will make you schizophrenic. Left-handedness increases your risk of cancer. Heck, if you believe journalists, half of the objects in the world give you cancer—cell phones, baby bottles, microwaves, NutraSweet, power lines.*

Good news: you can take off your tinfoil beanie. Most of these ideas are every bit as stupid as they sound. They are all the products of a particular kind of proofiness, an unwanted side effect that stems from the way our minds work.

We humans are incredible at recognizing patterns. Nothing in the world—not even the most powerful computer—is as adept as we are at spotting subtle relationships between objects or events. Finding patterns is deeply ingrained in our minds; after all, our very survival as a species depended on it.

If you get sick after eating a bit of shellfish, your brain almost automatically makes the association: shellfish causes sickness. The aversion you probably feel next time you come across a shrimp cocktail is an atavistic defense mechanism—your brain, having made the association between shellfish and poison, is trying to keep you safe by preventing you from eating it again. It's a pretty nifty trick; it only takes a single morsel of bad food before your brain gently starts telling you to steer clear of it. The faster our brains recognized important patterns—a certain rustling of branches is caused by a big animal hiding in the bush, a peculiar color to the sky heralds a dangerous storm on its way—the more likely we were to survive. Our minds are primed to find patterns, to find hidden

* The other half prevent cancer, of course. As columnist Ben Goldacre once put it, newspapers "will continue with the Sisyphean task of dividing all the inanimate objects in the world into the ones that either cause or cure cancer." Unfortunately, this is a fruitless endeavor, as some compounds, such as red wine, both cure *and* cause cancer, at least if you trust the headlines.

causes behind seemingly random events, because this is a mechanism that helps keep us alive long enough to reproduce.

The downside to this spectacular pattern-matching ability is that we go overboard, seeing patterns even when they're not there. We see portents everywhere. We convince ourselves that plane crashes happen in threes, that hemlines can foretell whether it'll be a good year or a bad year on Wall Street, and that the winner of a football game determines who will win the next presidential election. Sometimes we use patterns to try to alter the future, not just predict it. Watch any sporting event on television and you'll be treated to the players carrying out elaborate rituals to ensure victory through countless bizarre jinxes and mojos. Superstitions like these were born when a player thought he saw a pattern, a hidden cause behind certain events: a lucky pair of shorts would seem to make the team more likely to win a game, uttering a particular phrase seems to make someone more likely to strike out, or bouncing the ball exactly three times seems to make it easier to sink a free throw. But these patterns are all false—faulty beliefs brought on by our brain's overzealous attempts to make connections between events. Several powerful forms of proofiness are a product of this misfiring of our pattern-seeking behavior.

In 1992, future New Jersey congressman Rush Holt was the spokesperson for the Princeton Plasma Physics Laboratory, a high-tech facility devoted to fusion research, with an eye toward building power plants. Unfortunately, the future of fusion in the United States was getting increasingly grim; it seemed that budgets were about to get slashed before too long. As spokesperson, Holt had to try to justify to the public—and to legislators in charge of the budget—why the laboratory should consume tens and hundreds of millions of dollars in the quest for fusion energy.

Holt bolstered his case with several dramatic slides, scatter plots of data similar to the one below, showing how energy consumption (and production) benefits humanity. The data made a point quite clearly: the more energy a society consumes, the longer its citizens live, and the less frequently its infants die. The chart was the key to long life; if you want your citizens to survive longer, you should be building more power plants and, of course, pouring money into the Princeton Plasma Physics Laboratory to research future sources of energy. (Fund us if you want your children to live!)

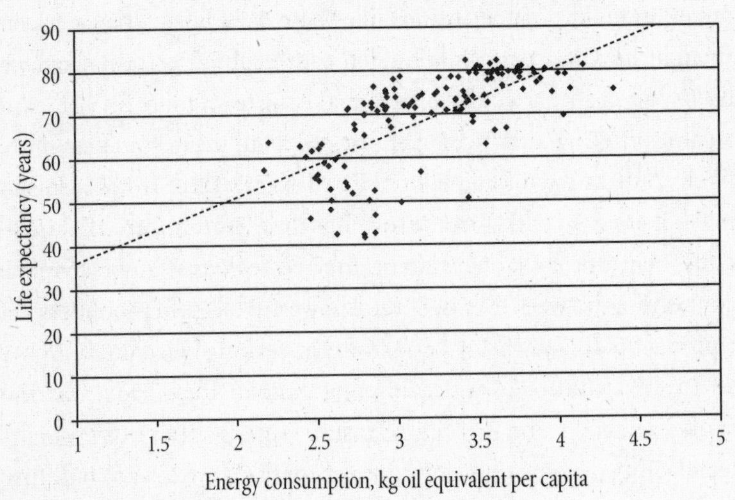

Figure 3. High energy consumption means a long life.

The data were correct, and the trend was real. The more energy people in a society produce and consume, the longer those people will live. More power equals longer life, right? Not so fast.

Holt's graphs showed that there was a tight relationship—a "correlation"—between power consumption and life expectancy;

the higher the power consumption, the higher the life expectancy. However, it's a classic mistake to say that you can increase life expectancy by increasing power consumption—to say power consumption causes longer lives.

Yes, it's true that the more power a society uses, the longer its citizens live, on average. It's equally true, however, that the more garbage a society produces, the longer its people live. The more automobiles people in a society drive, the more newspapers people in a society read, the more fast food people consume, the more television sets people have, the more time people spend on the Internet . . . in fact, the more edible underwear people in a nation eat, the longer the citizens of that nation will live, on average. Power plants don't lead to long life any more than garbage, Internet usage, newspapers, fast food, or edible underwear do.

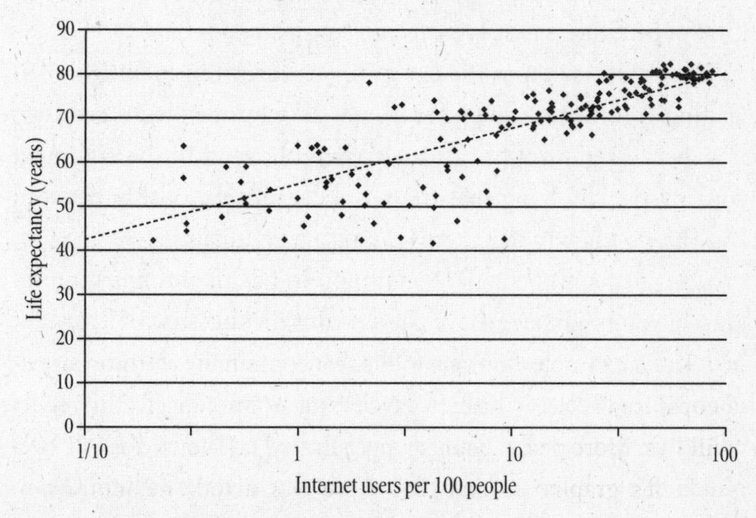

Figure 4. High Internet usage means a long life.

All of these data sets are correlated with each other. Power consumption, life expectancy, fast-food consumption, garbage production—all of these are large in a rich, industrial society and small in a poorer, agricultural one. At the same time, an industrial society tends to have modern hospitals and medicine, so its citizens tend to live longer and its infants tend to die less frequently. An industrial society also produces more garbage, spends more time on the Internet, eats more fast food, and, yes, needs more power than an agricultural society. So it's no surprise that these data are all correlated with each other—they all rise or fall together depending on how industrialized a particular society is. But this correlation is just that and nothing more; there's no "causal" relationship between edible underwear and infant mortality and television ownership. It was silly for Holt to imply that building more power plants raises a nation's life expectancy, just as it would be silly to suggest that we should be eating more fast food if we want to live to a healthy old age. Holt's presentation, in fact, was a vehicle for a kind of proofiness that I like to call *causuistry.*

Casuistry—without the extra "u"—is the art of making a misleading argument through seemingly sound principles. Causuistry is a specialized form of casuistry where the fault in the argument comes from implying that there is a causal relationship between two things when in fact there isn't any such linkage.

Causuistry is particularly common in health and nutrition research; you might even have altered your diet because of it. Lots of people, for example, don't eat foods that contain the artificial sweetener NutraSweet for fear of developing brain cancer. This belief comes from a bit of causuistry perpetrated in the mid-1990s by a bunch of psychiatrists led by Washington University's John Olney.

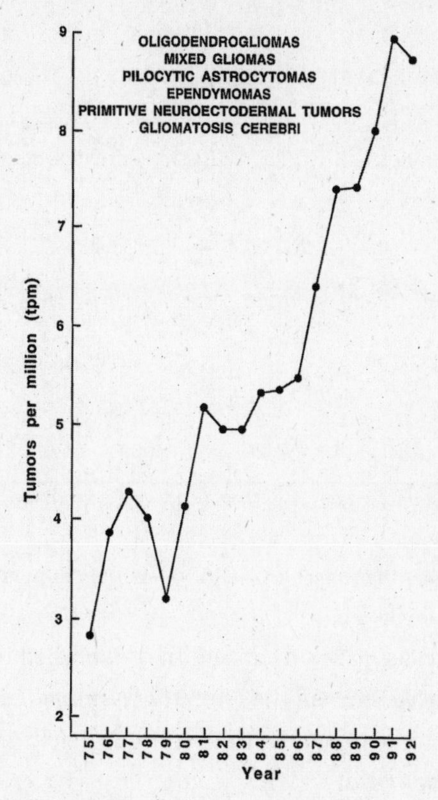

Figure 5. Brain tumors on the rise.*

These scientists noticed that there was an alarming rise in brain tumor rates about three to four years after NutraSweet was introduced in the market.

Aha! The psychiatrists quickly came to the obvious conclusion:

* Sharp-eyed readers will notice that the graph is apple-polished; just like the Quaker Oats graph in chapter 1, the scale of the graph is tweaked so that it doesn't start at zero as we assume. As a result, the rise in cancers looked much more dramatic than it actually was.

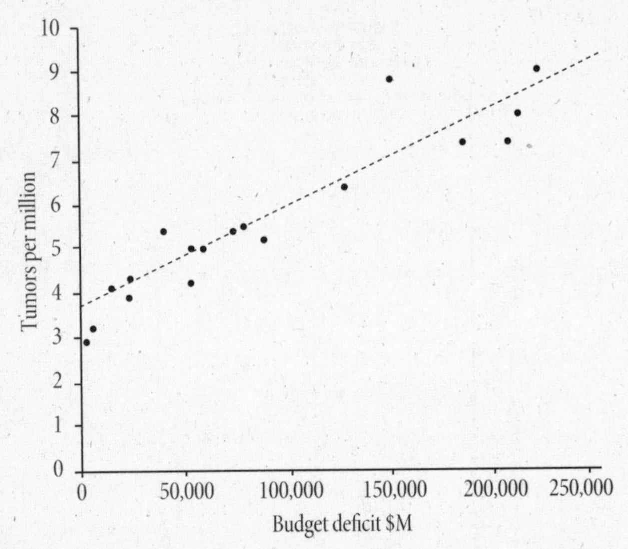

Figure 6. The higher the budget deficit, the more brain tumors there are.

NutraSweet is causing brain cancer! They published their findings in a peer-reviewed journal, the *Journal of Neuropathology and Experimental Neurology*, and their paper immediately grabbed headlines around the world.

But a closer look at the data shows how unconvincing the link really is. Sure, NutraSweet consumption was going up at the same time brain tumor rates were, but a lot of other things were on the rise too, such as cable TV, Sony Walkmen, Tom Cruise's career. When Ronald Reagan took office in 1981, government spending increased just as dramatically as brain tumor rates. If you plot deficit spending—the amount the government spent beyond its budget—against rising brain tumors, yet another correlation is amazingly clear. As one increases, the other does as well. Indeed, the data seem to show that deficit spending has a tighter relationship with brain cancer than NutraSweet consumption does! However, it would be utterly ridiculous to try to publish a scientific paper linking deficit spending to brain cancer.*

The correlation between government overspending and brain cancer is just as solid as the link between NutraSweet and brain cancer. Yet Olney's causuistry caused headlines across the world, and some people stopped drinking diet sodas out of fear. It's particularly galling because there was pretty convincing evidence that NutraSweet wasn't responsible for the increase in brain tumors. Olney almost certainly knew that brain tumor rates had stopped rising (they were even dropping a bit by the early 1990s) even though NutraSweet consumption was steadily increasing. This inconvenient fact didn't deter him from publishing his paper, nor did it stop anti-NutraSweet campaigners from using it to try to get the sweetener banned. Olney's work created a myth about NutraSweet that persists unto this day.

Legislators around the world still attempt to ban NutraSweet periodically, in part because of its supposed cancer-causing properties.† In 2005, a British member of Parliament tried to get "this carcinogenic substance to be banned from the UK food and drinks market altogether." A state representative in New Mexico tried to ban it in 2006, and one in Hawaii tried in 2009. The fight against NutraSweet—all based on a false belief—will almost certainly con-

---

* Which is why I did precisely that—in the *Journal of Neuropathology and Experimental Neurology*, of course. The editor in chief had a good sense of humor.

† Anti-NutraSweet lobbyists make equally flimsy assertions that NutraSweet causes other diseases as well, such as multiple sclerosis, lupus, seizures, and mental retardation. They also like to point out that a former CEO of the company that developed and manufactured NutraSweet was none other than George W. Bush's secretary of defense, Donald Rumsfeld.

tinue for years hence. Olney's causuistry created a belief that will take decades to erase.*

The question remains: if not NutraSweet, what was responsible for the rise in brain tumors? Nobody knows. A guess—and it's just that, a guess—is that we got a lot better at diagnosing brain tumors in the mid-1980s. In 1984, there were 108 MRI machines in the entire United States. These machines are particularly good at finding problems in soft tissue, like the brain. In 1985, the number more than tripled, increasing to 371. At the same time the government started approving MRI scans for Medicare patients. It's quite possible that the new diagnostic tool allowed doctors to catch more brain tumors all of a sudden. Just as it's important to take the changing value of a dollar into account when comparing spending over time, it's important to take doctors' changing diagnoses into account when looking at disease trends. (To be fair, Olney attempted to do this, but his analysis was unconvincing.)

A change in diagnostic criteria can look just like an epidemic. Doctors, for example, are much more likely to diagnose autism today than they were twenty or even ten years ago. As clinicians have become more aware of the condition, and as the criteria for diagnosis have changed several times since 1980, it's hard to tell if there's a real rise in autism or whether doctors now have a single name for something that was called a variety of things before, making for an easier diagnosis. In a 2002 study, researchers discovered that California showed a threefold rise in autism diagnoses over a seven-year period. However, over

* Of course, failing to prove that something is dangerous isn't equivalent to proving that something is safe, and Olney presented other evidence (such as a study in rats) to try to bolster his case. However, it's a good bet that NutraSweet (and many other items that have been blamed for brain cancers, such as cell phones) is not guilty of the accusations, in part, for reasons described below.

the same period there was a corresponding decrease in the diagnosis of undifferentiated "mental retardation"—suggesting that the rise is due to the labels that doctors put on a disease, rather than any change in the children who are being born. The jury's still out, but it's by no means clear that there really is an epidemic of autism.

The perception is enough—once people think there's an epidemic, they try to find something to blame. Causuistry is their primary weapon. With the perceived autism epidemic, many people point their finger at vaccines—either the measles-mumps-rubella (MMR) vaccine or a mercury-based preservative that was used in some vaccines. Activists such as former *Playboy* model Jenny McCarthy have launched a blistering causuistry-based attack that is causing parents to question whether they should get their children vaccinated. The antivaccine activists note a correlation: children develop autism after they're vaccinated. Yet there's no causation there; in fact, there's plenty of evidence to the contrary. (For example, the rising number of autism diagnoses that predate the MMR vaccine's introduction in 1988 argues against an MMR-autism link; the preservative-autism link is undermined by the fact that Denmark's autism incidence keeps rising even though the preservative was phased out in 1992.) Blaming vaccines for autism is nothing but causuistry.

It's always difficult to differentiate causes from effects; it's hard to look at a set of data and come up with convincing evidence that event A is causing effect B. Sometimes common sense tells us that A and B aren't directly related at all—yet the pages of peer-reviewed journals are filled with silly-sounding "discoveries" of linkages that wither under the harsh light of logic. In 1996, for example, sociologists argued that women with larger hip-to-waist ratios gave birth to a larger proportion of sons than daughters, perhaps explaining why men in many cultures find hourglass figures desirable. This shouldn't have passed a basic sniff test. In humans, the sex of a fetus is determined by whether the sperm that fertilizes the egg has an X or a Y chromosome; unless there are some very exotic genetic effects going on, the mother has no influence at all on the sex of her baby. So how could the mother's hip-to-waist ratio influence the gender of the child? It can't.

If there's any relationship at all between the mother's hip-to-waist ratio and the proportion of sons she has—which is rather dubious—there are some other ways it could come about. Perhaps giving birth to boys (who tend to have bigger heads) stretches the pelvic ligaments more than giving birth to girls, leading to a larger post-birth increase in the mother's hip-to-waist ratio. It's a bit far-fetched to think that giving birth to boys can alter a woman's hip-to-waist ratio, but it's a heck of a lot less far-fetched than the reverse. If the effect is real, then it's probable that the researchers accidentally mistook cause for effect.

It's surprisingly common for researchers to do this—to declare that A causes B when in fact it's more likely that B is causing A. The debt-causes-bad-health study, which appeared in the journal *Social Science and Medicine*, is a good example, because the researchers had the causality pointer in exactly the wrong direction. When the authors found that the higher a person's credit card debt, the worse his health, they promptly concluded that bad debt was causing poor health. Perhaps, they argued, the stress associated with financial trouble was causing illness. However, it's well-known that people in bad health are in worse financial straits than healthy people; they have to pay medical bills, and illnesses can interfere with people's ability to earn a living. (You can see this connection most clearly with bankruptcies, which are often—more than half the time, according to one study—in families where members had serious medical problems.) The researchers who found the correlation between debt and health apparently forgot to consider that they had cause and effect bass-ackward—and instantly rushed to the wrong conclusion.

Causuistry like this affects public policy, especially in cases where cause and effect are difficult to disentangle. Drug policy in the United States, for example, is a mess, and the government is not above using a little bit of causuistry to scare parents about the dangers of illegal substances. A few years ago, a terrifying advertisement circulated in magazines. Signed by the Office of National Drug Control Policy, the American Psychiatric Association, and a number of other organizations, the ad seemed to speak from authority. "Marijuana and Your Teen's Mental Health," it blared. "Depression. Suicidal Thoughts. Schizophrenia." The ad continued:

> Did you know that young people who use marijuana weekly have double the risk of depression later in life? And that teens aged 12 to 17 who smoke marijuana weekly are three times more likely than non-users to have suicidal thoughts?
>
> And if that's not bad enough, marijuana use has been linked to increased risk for schizophrenia in later years.

The message is clear: if your kids smoke marijuana, they run the risk of going crazy—they'll catch the reefer madness!

The problem is that the link between drugs and mental illness is a really tough one to understand. It is plausible (and there's some evidence) that marijuana might trigger schizophrenia in some people, or perhaps exacerbate existing cases. However, there's a large amount of evidence that people with mental illnesses—schizophrenia, bipolar disorder, depression, and other conditions—turn to drugs (and alcohol) as a means of lessening their symptoms, a phenomenon known as self-medication. However, "schizophrenia might make your kids smoke marijuana" is not nearly as scary as "marijuana might make your kids schizophrenic." It would never work in an advertisement. For different reasons, neither would the statement "alcohol might make your kids schizophrenic," even though alcohol's relationship to mental illness is similar to marijuana's.

Alcohol causes all sorts of problems for anti-marijuana forces. It has the same correlation to mental illness—and to taking hard drugs—as marijuana does. And if you press drug-policy causuists about alcohol, they might even claim that it's similarly dangerous. In the mid-1990s, midway into Bill Clinton's administration, there was a big push to classify alcohol and tobacco, along with marijuana, as "gateway drugs": substances that lead toward the use of nastier mind-altering substances like cocaine and heroin. That push—particularly the anti-tobacco element—got a lot of momentum from studies like the one performed by the Center on Addiction and Substance Abuse in 1994. The study showed strong correlations between tobacco and drug use. Kids who smoked tobacco were roughly fifty times more likely to snort cocaine and twelve times more likely to use heroin than their nonsmoking peers. The implication was clear: smoke a Marlboro and you're on your way to being a drug-addled junkie. The head of the addiction center, Joseph Califano, told the Senate, under oath, that tobacco was the "drug of entry into the world of hard drugs." His solution? A tobacco tax. Two dollars or more per pack of cigarettes to keep our children from buying them—it's a small price to pay to prevent our precious kids from becoming heroin addicts and cocaine fiends.

As big a public health problem as cigarettes are, they're almost certainly not turning our kids into junkies. The kinds of personalities that can lead to drug use—a fondness for taking risks, a sensitivity to social pressure—are probably the same ones that lead kids to try cigarettes. It's not that the smoking was causing the later hard drug use; it's just that smoking and drug use might stem from the same cause. Yet Califano was more than happy to make the logical leap and claim that tobacco leads children down the path to harder drugs—and to use this causuistry as a bludgeon to push for a tobacco tax.

Not so coincidentally, this tax was supported by Democrats. Big tobacco had donated more than five million dollars to Republican campaigners (much more than they donated to Democrats), thus attacking tobacco companies was good for the Democrats and bad for Republicans.* The Democrats used whatever weapon they could to hamstring donors to their opponents' cause, regardless of whether the logic behind the attack was sound or not. (Politicians seldom let facts get in the way of attacking their opponents.) Even though the tobacco-as-gateway-drug campaign was mere proofiness, it was a useful political tool.

Causuistry stems from our need to link every effect to a cause of some sort. Anytime we see two things that are correlated in some way, our brains leap to the conclusion that one thing must cause the other. It's often not so. Sometimes we mistake cause for effect and vice versa. Sometimes the cause is hidden away out of reach. And

* As ridiculous as the tobacco-causes-heroin-use causuistry was, it forced Republicans into the even sillier position that cigarettes were just dandy. "There is a mixed view among scientists and doctors whether it's addictive or not," Republican senator and presidential candidate Bob Dole said in 1996. Dole's stand on tobacco was, shall we say, quixotic. "We know [smoking is] not good for kids," he had said a few days earlier. "But a lot of other things aren't good. Some would say milk's not good."

sometimes the cause doesn't exist at all. That is perhaps the most difficult idea for humans to wrap our minds around: sometimes things happen for no apparent reason, just out of sheer random chance. If there's one concept that we humans have a hard time understanding, it's randomness.

Our minds revolt at the idea of randomness. Even when a set of data or an image is entirely chaotic, even when there's no underlying order to be found, we still try to construct a framework, a pattern, through which we understand our observations. We see the haphazard speckling of stars in the sky and group them together into constellations. We see the image of the Virgin Mary in a tortilla or the visage of Mother Teresa in a cinnamon bun.* Our minds, trying to make order out of chaos, play tricks on us.

Casinos make so much money because they exploit this failure of our brains. It's what keeps us gambling. If you watch a busy roulette table or a game of craps, you'll almost invariably see someone who's on a "lucky streak"—someone who has won several rolls in a row. Because he's winning, his brain sees a pattern and thinks that the winning streak will continue, so he keeps gambling. You'll also probably see someone who keeps gambling because he's been losing. The loser's brain presents a different pattern—that he's due for a winning streak. The poor sap keeps gambling for fear of missing out. Our minds seize on any brief run of good or bad luck and give it significance by thinking that it heralds a pattern to be exploited.

* The "nun bun," discovered in Tennessee, disappeared in 2005. It made a brief appearance in Seattle two years later, but as of the time this book was written, the bun's on the run.

Unfortunately, the randomness of the dice and of the slot machine ensure that there's no reality to these patterns at all. Each roll of the die, each pull of the lever gives a result that is totally unrelated to the events that came before it. That's the definition of random: there's no relationship, no pattern there to be discovered. Yet our brains simply refuse to accept this fact. This is *randumbness*: insisting that there is order where there is only chaos—creating a pattern where there is none to see.

Randumbness has a powerful hold over us. It gets even very smart people to believe some idiotic things. In 2005, anthropologists published a study in *Nature*—arguably the most prestigious peer-reviewed journal in the world—that showed how randomness can make even the most ridiculous of ideas seem credible. The study was an analysis of the results of wrestling, boxing, and taekwondo matches in the 2004 Olympics. In these contests, one athlete always wore red and the other wore blue. It so happened that the people who wore red won roughly 55 percent of the time, beating their blue-clad colleague. The conclusion? Wearing red helps athletes win.

Just a moment's thought should tell you that this is a pretty absurd notion. How could the choice of a red or blue jersey give a significant advantage to an athlete? But the authors of the paper were able to construct a "just so" story that made the discovery seem respectable enough for publication in a premier scientific journal: they cited a paper where experimenters put red bands on zebra finches, causing the wee birdies to act more dominant. If it's true for finches, why can't it be true for humans? The referees at the journal bought it, it was published, and media around the world trumpeted the finding to their audiences.

In truth, the authors were simply looking at a bunch of random data and constructing a pattern out of that random event. It was a fluke that slightly more red-wearing athletes beat blue-wearing athletes than vice versa, and the anthropologists created a study to explain that fluke. (If they had found that there were more blue-wearing victors than red-wearing ones, they could have constructed an explanation by using a bird species, like the purple martin, where blue coloration is a sign of dominance.) However, it's clear that the researchers' result is sheer randumbness. If the anthropologists took the time to check other Olympics, they'd no doubt find that their seeming pattern disappears. Indeed, analyzing sports in the 2008 Olympics shows there's no advantage to wearing red; if you look at the winners of the same events, there seems to be a slight *disadvantage*. Athletes wearing blue won more matches than those wearing red, particularly when it came to freestyle wrestling. The "advantage" to wearing red simply evaporated with a new set of data.*

If you take any random collection of data and squint hard enough, you'll see a pattern of some sort. If you're clever, you can get other people to see the pattern too. For example, you can take a completely random collection of data and plot it; it'll look like a shotgun blast—there's no order to be found. However, if you take that shotgun blast and draw a line through it, you can make people see a pattern when there isn't one. If you do it convincingly enough, you might even get it published in a peer-reviewed journal. Diagrams such as the following, which appeared in *Nature* to support

* A failed prediction is a very solid sign that a pattern is phony. A pattern allows you to make a prediction; when you think that red gives an advantage over blue, you can predict that, on average, more red-wearing athletes will win an event, and you might even be able to make some money at the bookie's. A false pattern has no predictive power; it might seem to give you a lot of power to understand past data, but it completely breaks down when tested against new data. We'll see this again and again. (Yes, this is a pattern.)

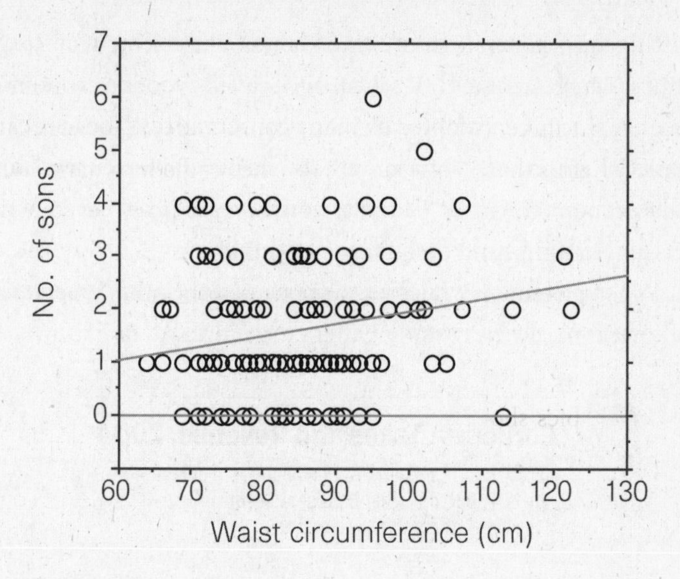

Figure 7. Faux order in chaotic data.

the idea that wide-hipped women give birth to more sons, would be a laughing matter if people didn't take them so seriously.

Drawing a line or curve through a clot of data is a very powerful method of shaping the way people interpret it. The line is a symbol of order; it shows that a pattern has been found within the raw scattershot chaos of points in the graph. Even if our eyes are unable to see the pattern directly, the line tells us what we *should* be seeing—even when it's not there.

For example, in 2007, an editorial in the *Wall Street Journal* slammed the United States for having such a high corporate tax rate—paradoxically, the editorial argued, cutting the tax rate would increase tax revenues. It's a counterintuitive position; if you're a government that wants to collect more money from your citizens, you should raise taxes, not lower them. However, the claim that cutting taxes generates more revenue was a key element of Reaganomics (which George H. W. Bush once called "voodoo economics") and has since been adopted by many conservatives. The idea can be depicted graphically by a mountain-shaped Laffer curve, named after economist Arthur Laffer: as you increase taxes, revenue rises, hits a maximum, and then falls to nothing.

When plotting tax rates versus revenues, the *Wall Street Journal* editorial board clearly saw a Laffer curve in the data.

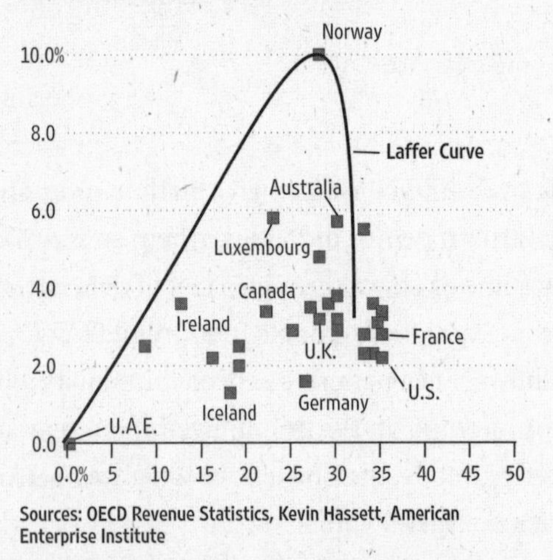

Figure 8. A laugher curve.

It's obvious that the *Wall Street Journal* would have drawn a leprechaun in the dots if that's what Reaganomics had predicted. There isn't any justification for drawing a Laffer-like curve that rises and falls within this particular data set; if anything, a straight line going up and to the right fits the data best. But by zipping the curve through an outlier—Norway—the *Journal* used randumbness to make Arthur Laffer look like a prophet even as the data made him look like a fool.

There's danger of proofiness even when data do seem to fall neatly along lines or curves, even when there seems to be a pattern nestled in the numbers. Just because a statistician or an economist or a scientist has discovered a real, bona fide relationship between sets of data doesn't mean that that relationship has any meaning. A line or curve on a graph, equation, or formula can represent a tight relationship among a vast amount of data—yet it might have no value whatsoever.

A good example of this is a 2004 *Nature* paper written by a motley collection of zoologists, geographers, and public health experts. This illustrious group of scientists analyzed athletes' Olympic performance on the 100-meter dash over the years and found some striking patterns. Male sprinters were getting faster and faster over the years; their times on the 100-meter dash were decreasing so steadily that you could draw a straight line through the data (which the researchers did). Female sprinters were also getting faster in a similar manner, also explained nicely by a line.

These graphs seemed to explain the data beautifully; the data never strayed far from the lines, so the researchers expressed high confidence that the lines described how men and women perform on the 100-meter dash, even far into the future. And if you follow those lines out, you see that they cross—women match and then surpass men—around the year 2156. The conclusion: women

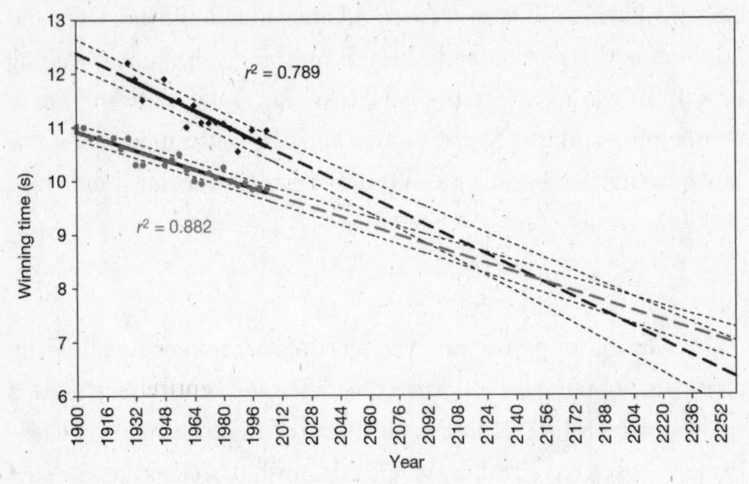

Figure 9. A graph that shows women outrunning men after 2156.

sprinters will be faster than men sometime in the middle of the next century.* After all, the lines fit the data, and when the lines cross, women outrace men.

However, those lines hide an absurdity. Follow them out for a while and the silliness becomes apparent. By the year 2224 or so, according to those lines, women will be running the 100-meter dash in seven seconds—a speed of roughly 32 miles per hour. Still plausible... barely. But the line keeps going. If you were to keep following it, you'd see that 150 years later, women would be sprinting at about 60 miles per hour. In roughly the year 2600, they would break the sound barrier. Shortly thereafter, they'd break the speed of light and travel backward in time, winning races before they actually begin. It's *impossible*

* Hedging their bets, the team did some complicated computer simulation that indicated that the "momentous" day when women beat men in the 100-meter dash could come as early as the year 2064 or as late as 2788.

that those lines truly represent what is going on in the data—they're faulty representations of reality.

Even though the lines seem very convincing at first glance, they're not truly capturing the underlying pattern in the data. Women have been racing competitively for a shorter time than men have, so it's no surprise that women's race times have been improving very rapidly, at least compared to men. As the sport matures, the improvement will slow down; the line will shift course, flattening out. Eventually, as both men and women reach their physiological limits, the improvement will stop entirely. The lines will be horizontal.

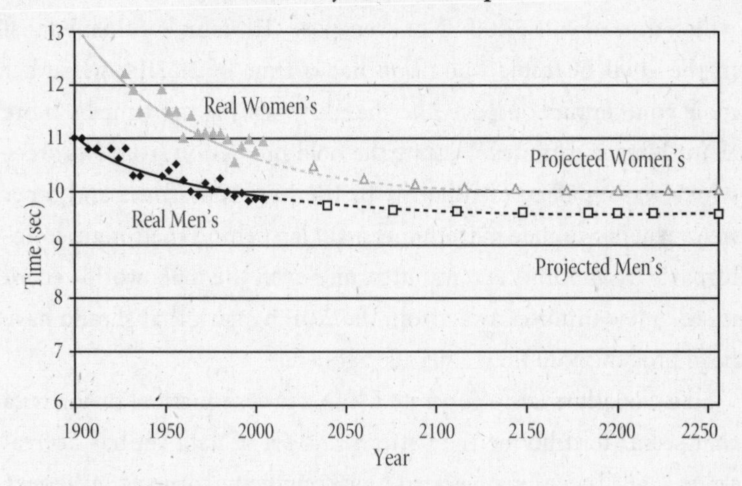

Figure 10. A more likely scenario for future sprints.

The lines need never cross at all; indeed, it's quite possible that the men's and women's sprint times will get closer for many years, but the women will never beat the men. What's certain, though, is that the data can't follow straight lines indefinitely. The laws of nature say so.

Mere natural law didn't stop scientists from making their bogus prediction, grabbing headlines with the prophecy that women will outsprint men in 2156. And it didn't stop *Nature* editors from accepting the paper. *Nature* should certainly have known better, and not just because of the inherent silliness of having women breaking the sound barrier in the middle of the millennium. In fact, the journal had been caught before with exactly the same sort of stupid prediction.

In 1992, two physiologists looked at women's and men's race times and blithely drew meaningless lines through the data. Their conclusion: female marathon runners would beat men after 1998, with a time of 2:01:59.00.* Not even close. The female gold medalist in the 2000 Olympic Marathon had a time of 2:23:14, while her male counterpart outpaced her handily, beating her time by more than thirteen minutes. Despite the bold prediction from *Nature*—which was repeated credulously by the *New York Times* and other newspapers—female marathoners still lag behind their male counterparts by about fifteen minutes, and even the male world record is still a few minutes away from the 2:01:59 mark that should have been broken more than a decade ago.

It's trivially easy to generate a line, curve, equation, or formula that seems to describe the pattern in a set of data yet has no real value at all. These faux patterns look convincing, dressed up as they are in mathematical language. But when people try to use them to

* Yes, they predicted the time down to a hundredth of a second—a clear-cut case of disestimation.

predict something—to exploit the pattern to say something new about the universe—they fall completely flat. Nevertheless, scientists, economists, public health experts, and all kinds of people with access to basic statistical software crank out meaningless curves, lines, equations, and formulae at an incredible pace. This is yet another form of proofiness: *regression to the moon.*

Regression analysis is a mathematical tool that people use to create lines, curves, formulae, or equations that fit a set of data. It's an extremely powerful technique; it quickly extracts a pattern from whatever data you provide it. However, when it is used incorrectly, the results are meaningless or downright barmy. The female-versus-male race times are obvious examples because the pattern that the scientists found in their data was inherently absurd; taken literally, it leads to the conclusion that runners will travel backward in time. Often the problems are more subtle.

In the 1980s, economists were ga-ga over a piece of research coming out of Yale University. A young economist, Ray Fair, had done a regression analysis on economic data from 1912 to 1976 and came up with an equation that, if correct, had stunning consequences—it predicted, well ahead of time, who would win a presidential election. All you had to do was plug in a few economic indicators—inflation, growth rate, and a few other factors—and voilà, out pops the next president. War, domestic issues, foreign policy—all of these were more or less irrelevant; the economic situation determines the winner of any given election. It's an equation that economists were sure to love.

Fair's equation predicted that Reagan would beat Carter with 55.4 percent of the votes cast for the two candidates. Reagan won with 55.3 percent. Economists around the world were thrilled. In

1984, Fair predicted that Reagan would trounce Mondale. He was right. In 1988, the equation said that George H. W. Bush would beat Dukakis handily. He did. Economists hailed Fair as a hero, and Fair preened, bragging about the accuracy of his predictions. His equation was taught in freshman economics classes as a tribute to the incredible power of economic forecasting. Then, in 1992, Fair declared that the winner of the upcoming election, with 57.2 percent of the vote, would be George H. W. Bush. He would annihilate the upstart William Jefferson Clinton, returning to the White House for a second term with an overwhelming mandate.

Pssssht . . . you could almost hear the economists' egos deflating when the election returns came in. Fair, humbled, tweaked his equation to correct for the mistakes he had made. In a paper published in mid-1996, just before the next presidential election, he gently ventured another prediction: "The basic story from the equation is that the 1996 election will be close with a slight edge for the Republicans." Whoops. Clinton defeated Dole almost as soundly, vote-wise, as Reagan beat Carter.

The problem is that Fair's equation was a regression to the moon. It was an elaborate model that found a pattern in the data, but the pattern was all but meaningless. (What success Fair did have boiled down to the commonsense dictum that incumbents benefit from a good economy.) The formula did a great job explaining past elections, but it was pretty hopeless when it came to predicting future ones: a sure sign of a faux pattern. Almost all electoral predictions have the same problems; year after year, economists and other experts line up with their regression models and make predictions that as often as not are dead wrong. On a slow news day, they even make it onto the front page of a major paper: "It's not even going to be close," an economist trumpeted from page A1 of the *Washington Post* in 2000—Gore would win 56.2 percent of the votes cast for the two main candidates.* Yeah, right.

The prize for regression silliness, though, has to go to the academics who crank out equations or formulae for everything under the sun, whether or not there's anything mathematically valid to make a formula about. It's a favorite pastime of attention-seeking pundits, as the media seem to gobble up these phony formulas without even a little bit of skepticism that might give them indigestion. In 2003, the BBC trumpeted a formula that philosophers have been seeking for years: the formula for happiness. This formula is simply:

$$\text{Happiness} = P + (5 \times E) + (3 \times H)$$

This equation is supposed to make sense when you know what the variables mean. P is "Personal Characteristics"—you get a high score if you have an optimistic outlook. E is "Existence" and reflects your health. H stands for your "Higher-Order Needs," such as the ego-stroking you get from fawning news organizations when you get them to publish your claptrap. The formula's obviously nonsense; these ideas are not quantifiable—there is no way to measure P or E or H—so it's a garbage equation that feeds on Potemkin numbers. This sort of equation is the lowest of all the varieties of

* Fed up with the prediction chatter, I wrote an op-ed in the *Post* that used an obscure numerical sequence, the 4-Knödel numbers, to predict the outcome of the election, which it had done accurately for every election since 1952. It was silly to say that such a sequence could determine the victor of the presidency, but it was no more silly—and no more wrong—than the prediction provided by economists with their regression to the moon.

regression to the moon; it isn't even a good-faith attempt to try to explain data. Yet these phony equations emerge from the swamps of academia quite regularly. Want to know the most depressing day of the year? Use the formula that came from Cardiff University:

$$\text{Misery} = 1/8W + (D - d)\ 3/8 \times TQ\ M \times NA$$

where W is weather, D is debt, M is motivation, NA is the need to take action, and so forth and so on.* In 2005, when the formula was first revealed, it proved—scientifically—that the most miserable day of the year was January 24.

There are many ways to generate numerical falsehoods from data, many ways to create proofiness from even valid measurements. Causuistry distorts the relationship between two sets of numbers. Randumbness creates patterns where none are to be found. Regression to the moon disguises nonsense in mathematical-looking lines or equations or formulae, making even the silliest ideas seem respectable. Such as the one described by this formula:

$$\text{Callipygianness} = (S + C) \times (B + F) / (T - V)$$

where S is shape, C is circularity, B is bounciness, F is firmness, T is texture, and V is waist-to-hip ratio. This formula was devised by a team of academic psychologists after many hours of serious research into the female derriere. Yes, indeed. This is supposed to be the formula for the perfect butt.

In fact, it's merely a formula for a perfect ass.

* For those of you who noticed that the formula, as presented, doesn't make mathematical sense . . . you are correct. It scarcely matters, though.

3

Risky Business

In each case, the companies and their executives grew rich by taking on excessive risk. In each case, the companies collapsed when these risks turned bad. And in each case, their executives are walking away with millions of dollars while taxpayers are stuck with billions of dollars in costs.

—Henry Waxman, Hearings on the Causes and Effects of the AIG Bailout, October 7, 2008

The chances of a disaster were stunningly large.

On July 18, 1969, two days before Neil Armstrong and Buzz Aldrin first set foot on the moon, a speechwriter in President Nixon's office penned a speech for an all too probable situation:

> Fate has ordained that the men who went to the moon to explore in peace will stay on the moon to rest in peace. These brave men, Neil Armstrong and Edwin Aldrin, know that there is no hope for their recovery.

> But they also know that there is hope for mankind in their sacrifice.
>
> These two men are laying down their lives in mankind's most noble goal: the search for truth and understanding. They will be mourned by their families and friends; they will be mourned by their nation; they will be mourned by the people of the world; they will be mourned by a Mother earth that dared send two of her sons into the unknown. In their exploration, they stirred the people of the world to feel as one; in their sacrifice, they bind more tightly the brotherhood of man. . . .
>
> For every human being who looks up at the moon in the nights to come will know that there is some corner of another world that is forever mankind.

It's hard to fathom just how risky the moon missions really were. In the early days of the Apollo program, NASA asked General Electric to calculate the chances of successfully landing men on the moon and bringing them back to earth in one piece. The answer they got was shocking: less than 5 percent. According to the (admittedly crude) numbers, a moon landing was so risky that it would be foolhardy to attempt. If NASA had paid attention to the calculations, the Apollo missions would have had to be scrapped—a horrible political disaster for NASA and a humiliation for the nation. So NASA did what NASA tends to do under such circumstances: it crumpled up the calculations, tossed them in the garbage can, and went ahead with the program anyway. In this particular case, it happened to be the right decision. Even though there was an extremely close call—the Apollo 13 mission came within a hair of killing three astronauts—

the moon landings would never have happened had NASA not disregarded the risks.

NASA engineers were famed for their "can do" bravado. No task was too difficult for the scrappy rocket scientists. The agency sneered at even the most daunting odds. Even if the numbers were dismal and the risks were enormous, NASA administrators disregarded them and plowed onward—with good reason. If Congress got wind of just how risky NASA's human spaceflight projects were, the programs could well have been canceled. It was in NASA's interest to disregard risks to keep their projects alive. On occasion, they even diddled with the numbers to make it look like their rockets were much safer than they actually were.

In 1983, the air force commissioned a study to calculate the risk that the brand-new space shuttle launch system would explode during launch. The study found that there was a dangerously high probability of disaster. As two of the study's authors wrote, "The probability of a solid rocket booster (SRB) failure destroying the shuttle was roughly 1 in 35 based on prior experience with this technology." One in thirty-five was an enormous and unacceptable level of risk. After all, the shuttles were supposed to make hundreds of flights, returning their crews safely every single time. If the shuttle would, on average, lose a crew of seven astronauts once every thirty-five flights, the shuttle program was as good as dead. So NASA disregarded the study, instead deciding "to rely upon its engineering judgment and to use 1 in 100,000 as the SRB failure probability estimate." In other words, NASA simply tossed out the 1-in-35 number and substituted a much more acceptable one—in which you could launch a shuttle every day for decades, totaling thousands upon thousands of launches, and expect not to have a single failure.

On January 28, 1986, a bit more than half a second after *Challenger* left the launch pad, a puff of gray smoke coming from its right solid rocket booster heralded disaster. Nobody knew it at the time, but a small rubber seal in the booster had failed. Fifty-nine seconds into the flight, a small flame erupted from the booster and the conflagration quickly grew out of control. Seventy-three seconds after launch, at an altitude of 46,000 feet, *Challenger* exploded in an enormous yellow-white ball of fire. It had taken only twenty-five shuttle launches before the risks caught up with NASA.

NASA management had deliberately understated the risks of a shuttle flight. Instead of facing the unpleasant reality that the shuttle boosters were risky, the agency decided to engineer a lie that was more acceptable. As physicist Richard Feynman, a member of the *Challenger* investigation panel, put it, "As far as I can tell, 'engineering judgment' means that they're just going to make up numbers!" Instead of performing a genuine assessment of the probability that the shuttle would fail, the management would start with a level of risk that was acceptable and work backward. "It was clear that the numbers . . . were chosen so that when you add everything together, you get 1 in 100,000," Feynman wrote. NASA's risk estimates were complete fictions, and nobody noticed until disaster struck.

Risks are tricky. We're pretty bad at estimating them. We spend our time worrying about graphic but uncommon events (meteor strikes, child abductions, and shark attacks) when we should really be worrying about—and preventing—more mundane risks (strokes, heart attacks, and diabetes) that are much more likely to cut our lives short. We spend our money chasing after faint hopes of winning the lottery or hitting pay dirt in a get-rich-quick scheme instead of paying off the credit cards that have a serious chance of driving us into ruin. We are terrified of dying in a plane crash but think nothing about speeding down the highway while talking on a cell phone. We don't have an internal gauge of what behaviors are truly dangerous and what aren't.

In the 1980s, economists Daniel Kahneman (who would later win the Nobel Prize) and Amos Tversky showed how irrational humans can be when confronted with risk. They presented test subjects with a scenario in which they had to make a difficult choice:

> Imagine that the U.S. is preparing for the outbreak of an unusual Asian disease, which is expected to kill 600 people. Two alternative programs to combat the disease have been proposed.

The two programs are very different—one is conservative, with a high probability of saving a small number of people, and one is risky, with a small probability of saving a large number of people. The subject has to make a choice about whether to choose the conservative or the risky strategy. But there was a twist. Kahneman and Tversky presented the exact same choice, but with slightly different wordings, to two separate groups of subjects. For the first group of subjects, the wording emphasized saving people from the disease; for the second, the phrasing dwelled on the victims of the disease rather than the survivors.

These differences in wording were purely cosmetic. Mathematically speaking, the two scenarios were identical. If people were logical creatures, the first group of subjects should make the exact same choices as the second group. Yet Kahneman and Tversky found that the wording of the programs made a tremendous difference—it was the phrasing, not the mathematics, that determined how people would behave. When the phrasing emphasized survivors over victims, subjects voted overwhelmingly—72 percent to 28 percent—to eschew the risky course of action, instead taking the conservative course that saves some patients with high certainty but lets others die. But when the wording spoke of victims rather than survivors, subjects suddenly became less risk-averse; they voted even more overwhelmingly—22 percent to 78 percent—to make a desperate gamble to save lives. The risks were exactly the same in both cases, but people, with their bad sense of risk, couldn't figure that out. The test subjects made their decisions based not on logic but upon how an authority presented the risks to them.

Minor changes in wording can easily make a huge risk seem worth taking or an insignificant risk seem dangerous. As a result, we are vulnerable to manipulation. We can't easily detect when someone is understating or exaggerating risks. We are prey to *risk mismanagement.*

Perhaps more than any other form of proofiness, risk mismanagement means big business. In NASA's case, it meant billions of dollars. By understating the risks of shuttle flights, they got funding from Congress to pursue a disastrous program—one that would kill fourteen astronauts for very little palpable gain. And other entrepreneurs are hoping to follow in NASA's footsteps. However, spaceflight is just a tiny part of the picture. Much of our economy revolves around risk—risk and money go hand in hand. There are entire industries devoted to managing and assessing risk; corporations have figured out how to make incredible amounts of money by measuring risk, packaging it, dividing it, and moving it around. And where there's money to be made, there's risk mismanagement. If you gaze deeply into the center of the ongoing financial crisis, you see risk mismanagement staring back at you. Risk mismanagement is crippling our economy, and along the way making a small handful of malefactors very, very wealthy.

NASA isn't the only game in town when it comes to getting people into space on the back of risk mismanagement. Airline magnate Richard Branson is hard at work trying to snooker private investors just as NASA snookered Congress. Branson is currently running a private spaceflight enterprise, "Virgin Galactic," which within its first five years of operation is supposedly going to launch an estimated three thousand passengers into space. Safely. "Virgin has a detailed understanding of what it takes to manage and operate complex transportation organizations . . . such as Virgin Atlantic Airlines and Virgin Trains which carry millions of passengers each year and have enjoyed superb safety records," brags the Virgin Galactic website. If you believe Virgin, spaceflight will be no riskier than a little jaunt on a private jet.

Hogwash. By comparing spaceflight to plane and train travel, Virgin is effectively underestimating the huge risks you take when you strap yourself to a rocket. It's a very dangerous task to pack enough energy into a cylinder to get you into space—and it's equally dangerous when, falling through the atmosphere, you get that energy back and have to dissipate it away in the form of heat. Throughout the history of spaceflight, about one in a hundred human-carrying rockets has killed its passengers, and that risk seems unlikely to change in the near future.

One chance in a hundred might not seem like so much, especially for the rare privilege of becoming an astronaut. But as far as risks go, it's extraordinarily high. For comparison, if today's U.S.

passenger aircraft had a similar failure rate, there would be roughly 275 U.S. plane crashes and 20,000 fatalities every day. A one in a hundred chance of dying every time you set foot on a plane would doom the airline industry; a 1 percent chance of death is simply too risky for any form of transportation to be commercially viable. If the historical failure rate holds, at Virgin Galactic's projected launch rate of one flight per week there would be only a one in three chance that Virgin Galactic goes for two years without a *Challenger*-type disaster. All in all, their chance of getting all three thousand people into space and back home again safely in this (hypothetical) scenario would be about half of 1 percent. People would almost certainly die, sooner rather than later. Even if the company survived the inevitable investigation and embarrassment, it would be just a matter of months before another explosion.

In my opinion, Branson is downplaying the risks, which has helped him convince more than 250 astronaut wannabes to put down $30 million worth of deposits on rides into space. He's also sold politicians and the public on his vision. In 2005, New Mexico politicians started spending tens of millions of dollars—and the governor promised to raise as much as $225 million—to build a spaceport. Two New Mexico counties even passed a sales tax to fund the project. As a smart businessman like Branson probably knows, downplaying risks can be very lucrative. In fact, there are two main ways to mismanage risk for fun and profit. Like NASA or Richard Branson, you can underestimate risks, making something look safer than it actually is. Conversely, you can take something that's mundane and exaggerate its risks, making it loom large in the public's imagination.

Journalists are particularly fond of the latter course. The scarier the story, the bigger the audience. Nothing sells like Armageddon.

Around the turn of the millennium, asteroid scares were all the rage. In 1998, the discovery of a large asteroid gave reporters the opportunity to grab readers with headlines like "October 26, 2028 could be our last day." In 2002, another asteroid gave rise to similar worries: "The world ends on Feb 1 2019 (possibly)." Journalists seemed undeterred when astronomers repeatedly told them that the risk of an actual collision was low. More observations, they said, would be able to pinpoint the asteroids' orbits with greater precision and a smack-up would almost certainly be ruled out—as indeed it was. But the attention-grabbing stories were too good to pass up. In fact, every doomsday scenario, no matter how far-fetched, is guaranteed to get at least some level of attention in the media. Every time physicists start up a new high-energy atom smasher—the Tevatron in 1985, the Relativistic Heavy Ion Collider in 2000, the Large Hadron Collider in 2008—the press chatters on about bizarre claims that the machine will destroy the earth or even the universe. The airwaves are alive with theories that the new machine will create tiny black holes that will swallow the earth, will create a variety of particle known as a "strangelet" that will destroy the planet, or will change the structure of space and time in a way that might annihilate the entire universe. (While no scientist will say that these scenarios are *impossible*—after all, scientific knowledge is uncertain and tentative—every mainstream scientist agrees that they are all very, very improbable.) Yet every time one of these new machines turns on, there are always headlines such as "Is the end nigh? Science experiment could swallow Earth, critics say" and "Physicists fear Big Bang machine could destroy Earth." Nobody seems to care about the science; everyone is fascinated with the prospect of Armageddon, no matter how remote it might be.

Fear sells so well that news organizations occasionally cross ethical lines to make something appear more risky than it actually is. On November 23, 1986, *60 Minutes* correspondent Ed Bradley reported on allegations that a particular model of car—the Audi 5000—was prone to sudden and unexpected jolts of acceleration even while the driver was pressing on the brake. It was a moving report; a tearful mother told of her horror as her Audi lurched out of control and ran over her son. It was also startling. At one point, the program showed a terrifying sequence that begins with an Audi idling at a dead stop. All of a sudden, the accelerator depresses by itself, and the car is suddenly zooming without a driver. Audi sales plummeted—from about 74,000 cars in 1985 down to a nadir of 12,000. After all, if Ed Bradley was correct, the automobiles were clearly death traps.

It was an illusion, and a relatively implausible one at that. The engine in the Audi 5000 was not strong enough to overpower the brakes—if the driver was pushing hard on the brake, the car would have stayed put even if the engine was going full throttle. (And in all of the accidents in question, the Audi's brakes were in working order.) When the National Highway Traffic Safety Administration studied the issue, they found that the unintended accelerations were caused by drivers pressing the accelerator pedal instead of the brake. In each case, the panicking driver, trying to stop the car, stomped on the accelerator harder and harder, making the Audi zoom out of control. (Which explains why Audi accident investigators occasionally found accelerator pedals that had been bent by the pressure of the driver's foot.) A Canadian investigation found the same thing. The Audis weren't defective; there was nothing mechanically wrong with them. But *60 Minutes* had the unexpected acceleration on tape! What about the eerie footage of the accelerator pedal pressing down on its own and the Audi suddenly zooming out of control? This is where standard *60 Minutes* fearmongering turns, in my opinion, into bona fide journalistic misconduct. The demonstration seems to have been rigged. A safety "expert" had apparently bored a hole in the Audi's transmission and forced high-pressure air in—it was this high pressure that caused the accelerator to depress "by itself." Bradley never hinted that the Audi had been doctored, and, to all appearances, *60 Minutes* had successfully tricked its audience into thinking that a risk existed when it was just a fiction.*

Exaggerating risks—and the fear these risks cause—is a potent tool, and not just for journalists. Politicians have long known that the public tends to support its leaders in times of crisis, so there's an incentive to make threats seem bigger than they are. According to some critics, the Homeland Security Advisory System—the color-coded five-level terror alert system created shortly after the September 11 attacks—is nothing more than an institutionalized way for the executive branch to manipulate the perceived risk of a terrorist attack. Though it's hard to divine the motivations of the Bush administration, there are some disturbing hints that the alert system was manipulated for political ends. For one thing, the alert level has *never* been relaxed below stage three ("yellow") since the September 11 attacks, indicating an "elevated" risk of terrorist attack. "Elevated" means "higher than usual," so it's completely nonsensical to declare that the everyday state of affairs exposes us to a higher-

---

* Unfortunately, this sort of hanky-panky isn't unique to *60 Minutes*. In 1992, *Dateline NBC* ran a spectacular segment about the dangers of GM trucks, showing one bursting into flames when struck in the side. There was in fact a design flaw, but *Dateline* had neglected to mention that to ignite the blaze, the TV show had strapped rockets to the truck's gas tank. Unlike *60 Minutes*, *Dateline* was forced to apologize for its actions. Without a hint of irony, Don Hewitt, *60 Minutes*' executive producer, used the occasion to tout his show's superior ethical standards: "If that had happened at *60 Minutes*," he told the *New York Times*, "I'd be looking for a job tomorrow."

than-usual risk of terrorist attack. It's impossible to have an eternally elevated risk, just as it's impossible for all the children in Lake Wobegon to be above average. Another hint comes from indications that political considerations occasionally played a role in raising and lowering the alert level. For example, in 2003, in the lead-up to the invasion of Iraq, the administration admitted relaxing the alert from stage two ("orange") to stage three ("yellow") not because of any change in the level of threat, but so they could elevate it again a few days before the war began. And finally, there's a bit of evidence that is very disturbing even though it should be taken with a big grain of salt. A 2004 study seemed to indicate that every time the government issued a terror warning, the president's approval rate spiked higher. The benefits of scaring the public are all too apparent.*

Nevertheless, if you want to make oodles of money, underestimating risks, not exaggerating them, is where the money is. Richard Branson's millions and NASA's billions are just a drop in the bucket. Savvy businesspeople have figured out how to make themselves very wealthy through risk mismanagement, and their misdeeds are so great that they put the world economy into enormous danger. An unprecedented international economic crisis was born in 2007 because of risk mismanagement.

In the world of finance, risk and reward are flip sides of the same coin. A safe investment—one with extremely little risk—will earn

* The United States is not alone in its mismanagement of terrorist risks. In mid-2009, the United Kingdom relaxed its five-point terrorist threat level from 2 ("Severe") to 3 ("Substantial"). It was the first time since the system was made public in 2006 that the alert level has been that low.

an investor very little money. To make big profits, you have to be willing to bear substantial risks; you have to accept that your gains might never materialize and that you might even lose the money you invested. Successful investors aren't people whose bets always pay off; Warren Buffett regularly loses money when some of the risks he takes don't pan out. Successful investors are the people who have a knack for squeezing the most reward out of the risks they're willing to take. They're the people who maximize the upside, the return on an investment, while keeping the downside, the risks they bear, to a minimum. Conversely, if you want people to give you money—to invest in you—you have to give them a large return with a minimum of risk. The less risk you seem to represent, the more money people are going to give you. This is why risk mismanagement is such big business. If you can understate the risks in your business, if you can hide them or shuffle them around or pawn them off on someone else, you can make billions.

In the late 1990s, the energy firm Enron was rolling in money. Its officers were millionaires many times over; at one point founder Ken Lay was worth $400 million. Enron was producing enormous profits for its investors, and quickly became a Wall Street darling. In 1995, *Fortune* magazine named it the most innovative company in America. In 1996, it won the title again—and in 1997, 1998, 1999, and 2000. But it was all a façade of risk mismanagement. Lay and his colleagues were making their money, in part, by moving risk from place to place in an attempt to shield it from outside scrutiny. They had created a whole slew of shell corporations (many of which, such as Chewco, Obi-1 Holdings, and Kenobe, had *Star Wars*–inspired names) to hide how risky an investment Enron really was. These corporations assumed much of Enron's debt, taking it off the books. As a result, Enron looked squeaky clean when it was in fact saddled with billions in debt—and was joined at the hip to dozens of dubious corporations that were on the verge of bankruptcy. When the risk mismanagement came to light, investors fled and the whole house of cards collapsed, taking along with it the savings of many investors who had no clue that Enron was a risky investment.*

If you move money around in clever enough ways, you can camouflage risks and make an absolute mint. Bernard Madoff was worth more than $800 million at his peak—by shuttling money from client to client, he managed to hide just how risky his investment firm really was (and that he was making himself rich by stealing his clients' money). Through his manipulations, Madoff managed to deflect attention for more than a decade—until the markets collapsed, suddenly exposing the fact that his coffers were empty. In making himself rich, he managed to lose more than $50 billion of his clients' money, earning him 150 years in prison.

The Ken Lays and the Bernie Madoffs of the world are able to make their money by hiding risk and moving it from place to place, deceiving their investors. When their criminal enterprises collapse,

* There are supposed to be checks to make sure that this kind of monkey business doesn't happen. Enron was audited by an outside firm, Arthur Andersen, that was supposed to ensure its basic accounting honesty. But such accounting firms have a tremendous conflict of interest; after all, they're paid by the client whose books they're inspecting. As a result, these accounting watchdogs either turn a blind eye to risk mismanagement or face the possibility of losing an important client. More than a decade before the Enron affair, a similar pattern emerged in the savings and loan scandal, which cost taxpayers billions. The example of Silverado Banking, a Texas savings and loan, is typical. As one article put it, "In 1985 Silverado's auditors, Ernst & Whinney, forced the thrift to report $20 million in losses because of problem loans. Silverado's managers weren't pleased with that result, so they got rid of Ernst & Whinney and hired Coopers & Lybrand, which took a more flexible view of the books. In 1986, Silverado reported $15 million in profits and the managers got $2.7 million in bonuses. A year and a half later the enterprise collapsed, at a cost to the government of some $1 billion."

the malefactors rightly get a lot of press attention as they are pilloried and prosecuted. However, as big as these frauds might seem, they're nothing compared to the risk mismanagement that's gnawing away at the heart of our economy. Madoff and Lay are rank amateurs in the proofiness game.

There are more subtle ways to make money off of risk. Just as risk can be hidden or moved from place to place, it can be divvied up and sold. Insurance companies are nothing more than a dumping ground for risk; for an appropriate fee, you can have them assume a risk for you. Afraid of the risk of a house fire? Of hitting someone with an automobile? Of a lawsuit? For enough money, you can convince an insurance company to indemnify you. It's a lucrative business—if you understand your risks.

As an example, imagine that you want to sell fire insurance. You need to have a good estimate for the risk of a house fire (say that there's about a 1 in 250 chance that a typical American home experiences a fire in a year) and how much damage the average house fire causes (say $20,000). These two pieces of information tell you how much money you'll be paying out in claims, and thus how much you need to charge for the insurance. Indeed, given these numbers, you can make a very nice living by charging your customers $100 a year in premiums.

Because the risk of a fire is 1 in 250, for every 10,000 clients you have, you can expect roughly forty of them to have a fire in a given year. Each of those forty fires costs you roughly $20,000—a grand total of $800,000 in damage that your company has to cover. But at the same time, your clients are paying you $1,000,000 in premiums, leaving you with a handsome profit of roughly $200,000. All you do is sit in the middle and watch the money pour in, making the occasional payoff. To make money in this way, though, you really have to know your risks. If your risk estimates are off, if there are sixty fires in a year instead of forty, you'll have to pay off $1,200,000 in claims—$200,000 more than you collect in premiums, which means that you'll go bankrupt. So you do all you can to keep the risk of fire in your client base—your "risk pool"—as low as possible. You might encourage them to install sprinkler systems. Or you might insure only nonsmokers, as they are at less risk of setting a fire than the general population. Your survival as an insurer depends on understanding your clients' risks and reducing them.

That's the theory anyhow. Reality, though, is a lot more complicated. Say you don't want to be bothered with all the paperwork and hassle of handling insurance claims and making payoffs to clients. If so, you can pass the risk on to a bigger company, just as your clients passed the risk to you. You could take the bundle of 10,000 insurance contracts and sell them to another firm for, say, $100,000. You make less money than you'd get if you handled claims yourself, but you no longer have to worry about paying out claims. It's pure profit for you. And the big firm is happy too, because it also makes money on the deal—$100,000. That is, it makes money if your risk estimate is correct. If your risk estimate is off and, say, sixty clients burn their homes down, the firm loses a lot of money on the deal ($300,000 to be precise). As for you, the extra fires don't bother you at all! You no longer have a stake in how many clients make a claim on the insurance contracts you sold them. Regardless of what happens to their homes, you still make a profit of $100,000. Even if there's an epidemic of house fires, you can laugh all the way to the bank; you sold a firm bad risk for a nice little profit.

You can probably see where this is going. The moment you sold your insurance contracts to another firm, you stopped caring whether those contracts were risky or not. Your income depends only on the number of clients you sign up, not on whether or not they burn down their houses and make claims. You no longer have any incentive to maintain a safe risk pool; just the opposite, in fact. If you sell only to nonsmokers, you have fewer clients to choose from, and fewer contracts to sell to the big insurance firm. Instead, you should sell to smokers as well as nonsmokers, the better to drive the number of contracts up. You'll make more money that way. Heck, you can make a profit by selling insurance to fire-eaters, people who try to deep-fry turkeys every Thanksgiving, and serial arsonists. So long as you dump the contracts to an unsuspecting firm before your clients immolate their houses, you'll make money. For the scheme to work, though, you mustn't let the big insurance firm know that you're contaminating the risk pool—you have to pretend that the bundle of contracts is low-risk, when in fact it's absolutely crawling with firebugs. If you do it right, money flows from the large insurance firm into your pocket, all thanks to your clever risk mismanagement.

Instead of "fire insurance," use "mortgages," and all of a sudden you've got a (slightly oversimplified) explanation of the subprime mortgage meltdown in 2007. The risk mismanagement involved selling mortgages to people whose income wasn't sufficient to support one. There are reports of brokers encouraging people to lie on their mortgage application (a federal crime) to allow them to get a nice fat mortgage; in some cases, people didn't even need to prove that they had an income at all. The recipients of these loans were just treading water financially, and were highly likely to default on their loans. This made these so-called subprime mortgages extremely high-risk, but if you were able to dump them on somebody else's balance sheet, you could make a killing.

While times were good and the housing market was rising,

everybody who traded in mortgages and other monetary instruments based upon those mortgages (such as "credit default swaps," which were, functionally, unregulated insurance contracts on these loans) made a fortune. Brokers, banks, and insurance companies passed around the bad risks like a hot potato, creating a complicated web of risk and debt. Every time the mortgage potato circled around, CEOs, corporate officers, managers, traders, and brokers earned themselves millions of dollars in bonuses. By pretending that these high-risk loans were actually low-risk, everybody was making themselves extraordinarily wealthy. But when the economy softened and the housing market began to fall, the good times suddenly came to a screeching halt. People began to default on their mortgages in droves; it's as if an arson craze hit insurance clients all at the same time. There was a giant sucking sound as the bad risk began gulping money away from the firms and other investors who held large amounts of these "toxic assets." Citigroup. Merrill Lynch. Bear Stearns. Lehman Brothers. Morgan Stanley. J. P. Morgan. Freddie Mac. Fannie Mae. AIG. It's hard to know just how much money evaporated—how much of the net worth of these companies was based upon understating the risk of these mortgages—but the damage is in the range of hundreds of billions or even trillions of dollars. This is a mind-boggling sum, not far off from the annual budget of the entire U.S. government. The result was a global economic catastrophe of unrivaled proportions.

During the good times, employees at AIG, Citigroup, and other companies made themselves very, very rich through risk mismanagement. They deliberately underestimated—indeed, ignored—the huge risks of these mortgage-based investments, trading them around and siphoning off money from one another's companies. And when the mortgages blew up, they got to keep their cash. The companies they worked for, though, were saddled with crushing debt, and looked ready to collapse. The government had to step in to save them, at the cost of trillions of taxpayer dollars.

Yet much of that bailout money would wind up in the pockets of those who caused the crisis in the first place. It was almost inevitable—it was an outcome determined by the way humans deal with risk.

The rules that govern the behavior of officers at AIG, Citigroup, and other malefactors in the worldwide economic crisis are the same ones that are causing us to chop down the rain forests, to fill our atmosphere with carbon dioxide, and to overfish the oceans. This phenomenon, known as the *tragedy of the commons*,* is everywhere—even at a friendly dinner.

Imagine that you're at a fancy restaurant with a bunch of acquaintances and you've all decided ahead of time to split the bill evenly among you. Even though the restaurant is pretty expensive, the prix fixe menu isn't too pricy. It gives you a tasty if modest meal, say, a chicken breast, along with a small side salad. It's a wee bit spartan, but it's a very good deal. The other temptations on the menu, such as the lobster, are much more appealing, but their prices are exorbitant. If everybody sticks to the prix fixe menu, the bill will be relatively small. When you divide up the bill, you and your acquaintances will all get a very good deal—a pleasant meal at a cheap price. Alas, that's probably not what happens. More typ-

* So named because the original example concerned the English custom of having a community grassland (the "commons") that livestock owners could use for grazing. As the story goes, in an attempt to capitalize on the free grazing land, people bought more and more livestock and overgrazed the pastures, rendering them useless.

ically, someone at the party has to be a spoilsport and order the lobster. Lobster-boy thinks he's pulling a fast one; since everybody divides the bill equally at the end of the meal, the enormous cost of his lobster is shared among the whole crowd. He personally gets the benefit of eating the expensive entrée, but in the process he drives up the cost of the meal for everybody. Sadly, once lobster-boy puts in his order, all bets are off. Everybody starts ordering the overpriced lobster; even a few people who previously ordered the prix fixe special will change their orders. Perhaps a handful of people show restraint and stick to the cheaper chicken, but it's of no avail. All the lobsters drive the price of the meal up to enormous heights. When the check comes and it's divided up, the cost is eye-popping. This is an example of the tragedy of the commons.

The tragedy of the commons occurs when an individual can take an action that benefits him, yet the negative consequences of that action are diffused—such as when they're divided among a large group of people or when they take a long time to materialize. In situations like these, people act selfishly, getting as much benefit as they can, but as a consequence, we're all worse off. In the restaurant example, lobster orderers benefit from getting the expensive entrée, but the cost is divided among the entire party, making what should have been an affordable, pleasant meal into an outrageously expensive fiasco. Most environmental problems are the result of a tragedy of the commons. Burning fossil fuels, for example, gives the user benefit of cheap, affordable energy, even though each power plant helps warm the planet a little bit—and the consequences of global warming are shared by everybody in the world.

Another example of the tragedy of the commons is all too familiar to urban planners. Since the 1960s, a number of cities around the world have tried to encourage cycling by making bicycles freely available to anyone who wants to use them. It's a lovely idea; even if you don't use the bicycles yourself, you benefit from the reduction in automobile traffic. Alas, most of the time, these efforts wind up failing because people abuse the bicycles. In 1960s Amsterdam, most of the free bicycles were stolen within weeks. People are no better nowadays. In 2009, Paris's free-bicycle scheme was struggling to survive because people were abusing the bicycles. You can find videos on YouTube where young toughs ride the bicycles down stairs and jump them in skate parks. Vandals destroyed quite a few—smashing them, burning them, and tossing them into the Seine. And many others were simply stolen. A year and a half after the start of the Parisian free-bicycle program, more than half of the 15,000 bicycles were gone. The destructive behavior of individuals destroyed a shared resource. Tragedy of the commons is an immutable fact about society. If the benefits of our actions are divorced, to some extent, from their negative consequences, we're going to take those actions—even if they lead to a very unfortunate outcome for everybody. The tragedy of the commons is a result of human nature, and the moment we ignore human nature we put ourselves at grave risk.

In financial markets, the equivalent concept is known as *moral hazard*—it's what you get when you cross the tragedy of the commons with risk mismanagement. When financial gain is divorced from the risks involved in making that gain, very bad things happen. The subprime mortgage meltdown was a great example of moral hazard in action. The individuals who manufactured and traded those mortgage-backed securities were making oodles of money, but they were insulated from the risks inherent to those mortgages. As a result they misbehaved, misrepresenting risks on a tremendous scale to make money. Even though their actions little by little undermined the security of their companies—and the stability of the world economy—the party wouldn't stop until the whole economy threatened to come crashing down.

When the risk finally caught up with the securities and the whole risk-mismanagement scheme collapsed, these individuals tended to hold on to their wealth. Some of them even managed to make lots of cash even after the meltdown, thanks to another incarnation of moral hazard.

AIG, a massive insurance company, was at the heart of the meltdown, with tens of billions of dollars in bad mortgage risk. By the time the economic crisis came about, AIG had swollen to gigantic proportions. It had a trillion dollars in assets, and a finger in almost every big financial institution in the nation. If it collapsed, it would send tremors through all the banks and the investment firms in the United States; it might even cause some of them to collapse as well. People would panic, and potentially our entire economic infrastructure could be shaken to rubble in a matter of days. Economists call this potential *systemic risk*, and it was the ace in AIG's hand—by using that risk, AIG could ensure that the government would keep shoveling money into its coffers.

Because the government couldn't let AIG collapse for fear of destroying the economy, company officials realized that they had carte blanche to go wild and take whatever risks they wanted. AIG officials were largely insulated from the consequences of their actions, because they suspected that their company wouldn't be allowed to fail no matter what risks they took. This gave AIG's managers the green light for risk mismanagement on an enormous scale. They could deceive their investors about the risks that they were taking, misuse their assets, siphon off money through bonuses, and gamble on risky investments in hopes of reaping a windfall. If things turned sour, the government would have to step in and bail them out. Congress had no choice but to subsidize AIG's risk-taking.

Things did, of course, turn sour with AIG's risky investments, and the government duly stepped in. On September 16, 2008, the feds bailed it out with an $85 billion loan. "Nothing made me more frustrated, more angry than having to intervene, particularly in a couple of cases where taking wild bets had forced these companies close to bankruptcy," Federal Reserve chairman Ben Bernanke admitted in mid-2009. But because of the fear of triggering an economic disaster, "I had to hold my nose and stop those firms from failing."

Unfortunately, the bailout didn't stop the misbehavior. After all, AIG was still too big to fail—so long as the government was afraid of systemic risk, AIG officials had no reason to act honorably and try to put the company on a sound financial footing. Within days of the bailout, AIG executives were spending nearly half a million dollars of the bailout money on a retreat to a posh resort in California, where they indulged themselves with spa treatments, lavish banquets, and of course, plenty of golf—$7,000 in greens fees. Lawmakers were livid. But that anger wasn't enough to stop them from giving the company an additional $38 billion in October and then another $40 billion in November. Edward Liddy, the CEO of AIG, promised to behave. "We are tightening our belt," he told a reporter in October. "Just as the American consumer, the American taxpayer is tightening their belt, we are doing the same thing. But we're not stopping at one notch; we're going three and four and five notches." Apparently, AIG belts are constructed somewhat differently, because within a few months, Liddy was explaining to Congress why he felt it necessary to use the bailout money to pay

AIG employees $160 million in "retention" bonuses. The top bonus was a whopping $6.4 million. Seventy-three people got more than $1 million each—and eleven of them had already left the company, so the "retention" bonus could hardly be expected to work as advertised. It was unbelievable misbehavior, embarrassing Congress and the new president.

All companies that are too big to fail—Citigroup, General Motors, Fannie Mae, Merrill Lynch—are swimming in moral hazard. Once they know that they won't be allowed to collapse, it's almost guaranteed that they will fill their own pockets while passing the consequences of their risky behavior on to the taxpayers.*

Every few years, tremendous cases of risk mismanagement extract money from average citizens and put it into the hands of the wealthy. Whether it's the savings and loan scandal or Enron or the subprime mortgage crisis, the end is always the same. The people who are willing to lie about risks make themselves very rich, and the taxpayer suffers the consequences. Even if one or two of the malefactors wind up in jail, there are always many more who made themselves much better off at others' expense and never suffered any serious consequences.

Risk management is the form of proofiness that hits the pocketbook most directly. However, other forms of proofiness can have consequences that are even more grave. They can undermine the press, deny us our vote, put us in jail, and sap the strength of our democracy. Mere thievery pales by comparison.

* As I was finishing up this manuscript, the hemorrhage showed no signs of ending. The newspapers were expressing outrage about how Citigroup had announced that it would give its twenty-five senior executives enormous bonuses, including one worth a mind-blowing $98 million, making AIG look tightfisted.

4

Poll Cats

Public opinion contains all kinds of falsity and truth, but it takes a great man to find the truth in it.

— G. W. F. Hegel

"Dewey Defeats Truman."

It's an iconic picture. Harry Truman, grinning with uncharacteristic glee, holds the *Chicago Daily Tribune* up for all to see. "That is one for the books," he gloated.

The headline was a colossal screw-up, even by newspaper standards. Truman had trounced Dewey, but the first edition of the Chicago paper declared otherwise—scrawled across its front page in the biggest, boldest type it had available.

It's not unusual for first editions to have errors, as they're usually rushed to press. The *Tribune*'s first edition was even more rushed than usual because of a printers' strike—it had to be put to bed in the early evening. This was unfortunate, especially on an election night; the deadline was before even the East Coast returns were in. Without these results in hand, the editors had to make a judgment call about what to put in the election story on the front page. They chose to make a guess about the victor—a choice that was spectacularly risky and spectacularly wrong.

It didn't seem so risky at the time. All of the major polling agencies, including the big ones run by George Gallup and Elmo Roper, had long since concluded that Dewey would walk away with an easy victory. The results of their polls were so definitive that the pollsters closed up shop weeks before the election. Since the outcome was obvious, they reasoned, there was no need to continue collecting data. The *Tribune* editors, confidence buoyed by the incorrect polls, figured that Dewey was a shoo-in. But the polling experts had come to exactly the wrong conclusion.

What went wrong? According to Bud Roper, son of Elmo, nothing at all. "I think the 1948 polls were more accurate than the 1948 election." The polls weren't wrong—the voters were. This is idiocy of the highest order. Even in the face of overwhelming proof that they've screwed up, pollsters somehow still have undiminished faith in the phony numbers they generate.

Polls are perhaps the leading source of proofiness in modern society. They are an indispensable tool for journalists; it's hard to pick up a paper, listen to a newscast, or browse the Web for news without stubbing your toe on a poll—usually a ridiculous one—in a matter of minutes.* To politicians, polls are a tool that can help shape opinions, and a weapon that can help them attack their enemies. In the right hands, a carefully designed and executed poll

* One I encountered this morning: "Jesus and Princess Diana lead poll of dead people we most want to meet."

can give us an accurate snapshot of the collective thoughts of a society. Most of the time, though, a poll is a factory of authoritative-sounding nonsense—of proofiness.

The poll was a journalistic invention. Most historians say that the first one, at least in the political sphere, was conducted by the *Pennsylvanian*, a Harrisburg newspaper, in July 1824. A presidential election was a few months away, and the newspaper dispatched a correspondent to neighboring Delaware to see which way the political winds were blowing.* The answer came back: Andrew Jackson was greatly preferred over his rival, John Quincy Adams. It was a crude and flawed poll, but from this humble start, polling was born. Marketers, keen to find out what consumers think of their advertisements and their products, refined the art of the poll to a high level of sophistication. Politicians, like marketers, use polls to figure out what the public desires and to fine-tune their images accordingly. However, from the very beginning, the journalist was the prime purveyor of polls. He still is. To a journalist, a poll is a powerful mechanism for breaking out of shackles that subtly bind him.

Most journalists are primarily event-gatherers, picking and packaging the choicest and freshest events to present to their audiences. Every time there is a sufficiently interesting or important event of some sort—a plane crash, say, or an earthquake—journalists rush in to relay the story. However, without an event to report, journalists are almost helpless. When there's no event, almost by definition,

* This might be the etymology of "straw poll," from the method of holding up (or dropping) straws to get a better handle on the prevailing winds.

there's no news for them to report. As journalist Walter Lippmann put it in the 1920s:

> It may be the act of going into bankruptcy, it may be a fire, a collision, an assault, a riot, an arrest, a denunciation, the introduction of a bill, a speech, a vote, a meeting, the expressed opinion of a well known citizen, an editorial in a newspaper, a sale, a wage-schedule, a price change, the proposal to build a bridge. . . . There must be a manifestation. The course of events must assume a certain definable shape, and until it is in a phase where some aspect is an accomplished fact, news does not separate itself from the ocean of possible truth.

To a journalist, the event is a tyrant. It is the authority that grants him liberty to speak. And this liberty is typically only given for a short amount of time. Unless the event is extraordinarily salacious or deadly or important, the journalist must move on to other topics quickly, as his powers to attract an audience rapidly wane as the event ages. He has a day or two or three to talk about an explosion or child abductions before he must once more hold his tongue, at least until the next event.

To a reporter who's bubbling with ideas to write about, this can be terribly frustrating. Lots of interesting and important developments happen as a gradual trickle, rather than in a series of discrete, reportable events. However, journalists generally can't write about broad trends or abstract ideas until they find what is called a "news peg"—a timely event that the reporter can tie, no matter how tenuously, to the subject that he really wants to talk about. For example, a journalist who has a vague hankering to write about his suspicions that airline safety has been getting worse would keep an eye out for a news peg of some sort—any event that might provide a convenient excuse for publishing the story. A high-profile plane crash would be an ideal peg, but other lesser events—perhaps not newsworthy on their own—would also suffice. A near miss would do. So would an incident where a pilot gets fired for showing up on the job drunk. Reports are also good news pegs; the journalist probably wouldn't have to wait long before the FAA or some other government agency publishes a report or generates a new statistic about transportation that might imbue the piece with timeliness. Failing that, there's always an anniversary of some disaster or another; if desperate, the reporter can dust off TWA 800 or the Andes plane crash or even the R101 airship disaster to write a piece at the appropriate time. For a news peg need not even be a real event; it can be a fake one.

A real event tends to be spontaneous rather than planned; news happens on its own timetable. Even if the event isn't a complete surprise (everybody knows that an election is coming, for example), its outcome is at least somewhat unpredictable. A real event can be complex; it might take months or years to tease out its significance and it might never be understood fully. A fake event—what historian Daniel Boorstin dubbed a "pseudoevent"—tends to be just the opposite. Where real news is organic, pseudoevents are synthetic. A pseudoevent is planned rather than spontaneous. It occurs at a convenient time and at an accessible location. Any unpredictability is kept to a minimum. A good pseudoevent is simple and easy to understand. And it has a purpose. A pseudoevent like the presentation of a political speech or the orchestrated "leak" of a governmental memo is meant, at least in part, specifically for the consumption of the press—and once given an airing by the press, it is meant to get attention, to be talked about, and to shape public opinion. Though a pseudoevent might have information, that information has been massaged and molded with a purpose in mind. A plane crash has no hidden agenda; a speech from the president of Airbus certainly does.

Reporters make little distinction, if any, between events and pseudoevents. Both are useful; pseudoevents can serve as perfectly serviceable news pegs when an event is not readily available. A speech from the Airbus president can unshackle a reporter, allowing him to riff on the safety of airlines. Reporters are grateful for the freedom that the pseudoevents buy them, even though that freedom comes at the price of being manipulated by the creator of the pseudoevent. As a result, many corporations and government organizations have become adept at manufacturing pseudoevents that quickly get turned into pseudonews.

From the journalist's point of view, the poll is the ultimate pseudoevent—it is entirely under his control. Any time a news organization wishes, it can conduct or commission a poll, whose results it then duly reports. A poll frees journalists from having to wait for news to happen or for others to manufacture pseudoevents for them. Polls allow a news organization to manufacture its own news. It's incredibly liberating.*

What's more, polls allow reporters to bend real events to a convenient timetable, completely freeing them from the less than ideal

---

* Polls aren't the only way for news organizations to synthesize news. *Time*'s annual Person of the Year issue is a long-running exercise in pseudo-newsy attention grabbing. Top-ten and top-hundred lists are also very effective—and they seem to be proliferating rapidly.

timing of bona fide news events. During the doldrums of an election season, in the boring stretch when a vote might be weeks or months away, it might seem that news organizations wouldn't be able to talk about the election for lack of any events to report on. Not so. News organizations need only commission polls to give their reporters and talking heads something to pontificate about. Journalists chatter continuously throughout election season as if they were calling a horse race. Pundits spend countless hours rooting through the entrails of whatever national or local polls they can get their hands on, turning each little insignificant result into an important portent of future events. These polls allow the news media to keep their audiences tense and entertained even while crossing the vast, lonely electoral desert in between the results of the primaries—which usually aren't that interesting to begin with—and the general election in November. And as election day comes nigh, the polling gets even more intense. In days of yore, reporters had to wait until the returns were in before announcing the winner of an election. No longer! Exit polls allow the networks to declare a winner before bedtime. Polls are an incredibly powerful tool, and they've become a staple of modern journalism—and not just during election season.

This behavior reflects a great deal of journalistic faith in the reliability of polls. Journalists are, fundamentally, purveyors of nonfiction. So for them to use polls so extensively, they have to convince themselves that the polls are a reflection of external reality, that they represent truth, at least to some extent. This is why journalists don't regularly put astrological predictions in their copy, even though there's almost certainly a large audience that would be interested in what the stars hold for presidential candidates. Poll-based prognostications are trumpeted on the front page; astrological ones are relegated to the funny pages.* Even when polls fail—as they do, over and over again—the media's faith is undiminished and unquestioning.

That faith is symbolized by a little number that accompanies every major poll. Journalists interpret this number as a conditional guarantee of a poll's truth—it encapsulates all the uncertainty about a poll's veracity and rolls it into a tiny little ball that can be ignored. This number is arguably the most misunderstood and abused mathematical concept that journalists have gotten their fingers on: the *margin of error.*

Ask a bunch of journalists what a margin of error is, and you'll get a whole lot of contradictory answers. However, reporters tend to treat the margin of error as a litmus test about whether or not to believe the results of a poll.

For example, if a poll shows that 51 percent of the public would vote for George W. Bush and 49 percent would vote for John Kerry, it looks like a very slim two-point lead for Bush. No doubt many pundits would describe the result that way. However, the more cautious journalists in the newsroom wouldn't accept that result blindly. They would compare the difference between the two candidates to the poll's margin of error, which, for reasons to be explained shortly,

* Most of the time. Every election, a handful of major news outlets report astrological predictions. In mid-2008, ABC reported that astrologers predicted that Barack Obama "would win the White House in November" because Saturn was in opposition to Uranus. In 2004, Reuters quoted an Indian astrologer: "It is cosmic writ that George W. Bush cannot become President of the United States again." These stories are often, but not always, banished to "News of the Bizarre"–type sections where journalists titillate their audiences by flirting with outright falsehoods.

typically hovers around 3 percent. Once they discover that the lead—2 percent—is less than the margin of error—3 percent—they typically mumble an incantation about statistical significance and reject the lead, declaring the poll to be a dead heat. If, on the other hand, the difference between the two candidates is greater than the margin of error, then the lead is real: the poll's result is to be believed. Journalists use the margin of error as a touchstone, as a nearly infallible guide that allows them to separate good polls from bad; valid data from dross. Journalists' faith in this test is somewhat surprising, given how few of them understand what a margin of error really represents.

The margin of error is a rather subtle concept that describes a kind of problem that plagues polls. At its core, a poll is a measuring instrument designed to gauge public opinion, and, like any other measuring instrument, a poll is imperfect. Polls are inherently error-prone. However, there's a peculiar type of error that stems from the way that a poll is conducted—one that springs from randomness.

To make a measurement with a poll, a pollster asks a set of people—the "sample"—some questions. From the answers, the pollster knows, at least in theory, what the sample believes about a certain burning issue. With that data, the pollster then makes a leap of faith. The pollster assumes that the preferences of that small sample of people accurately represent the predilections of the entire population—the "universe." That is, if the pollster queries a sample of, say, a thousand people in Britain and finds that 64 percent prefer tea to coffee, she concludes that roughly the same thing is true of the entire universe of British citizens: she reports to an eager public that roughly 64 percent of Britons prefer tea to coffee. Unfortunately, this leap of faith introduces an unavoidable error. Even under the very best of circumstances, even when the pollster is extremely careful about selecting a sample of people to poll, the randomness of nature conspires to prevent her from getting a precise answer.

This error makes sense if you think about it; there has to be some kind of error associated with the size of the sample in a poll: the smaller the sample, the bigger the error and the less reliable it will be. A sample size of one is all but useless—asking a single random stranger what he likes to drink is not going to give you a fair cross section of what all of Britain enjoys. You might have randomly stumbled across a loon who prefers drinking sheep's blood to tea or coffee. In that case, your sample would indicate that Britain's number one drink is sheep's blood; the oddball result completely throws off the results of your poll.

The more people in your sample, the less likely you are to have weird random encounters like this that invalidate your data. If you poll ten people instead of one, a single sheep's-blood fanatic won't throw off the results as dramatically. Better yet, if you survey a hundred people, a single oddball will have a negligible effect on the outcome of the poll. So the larger the sample, the less error there is due to random weirdness.

In fact, random weirdness is a law of nature; no matter how careful you are in picking your sample, you're going to get strange events. Since this random weirdness throws off the accuracy of your poll, it is a source of imprecision dubbed *statistical error*. Luckily, there are mathematical and statistical equations that quantify this weirdness—they dictate what level of weirdness (and thus statistical error) you expect to get in a sample of a given size. In other words, there's a strict relationship between sample size and statistical error. The bigger the sample, the smaller the statistical error.* And though this might seem like hand-waving mumbo-jumbo, statistical error is a very real and important phenomenon that is grounded in fundamental mathematical laws of probability. Statistical error is a verifiable source of error in every poll, and it's a consequence of the randomness of nature—and a function of the size of the sample.†

The concept of margin of error relays this bizarre imprecision caused by randomness. By convention, it is a number larger than the imprecision caused by randomness 95 percent of the time. This is almost impossible to understand at first, so let's use an example.

Say that a poll finds that 64 percent of Britons prefer tea to coffee. The pollster knows that the randomness of the universe—statistical error—might mess up the result of the poll; the real answer might not actually be 64 percent, but instead 62 percent or 66 percent or even 93 percent if there was a particularly weird random event that messed up the sample. When the pollster says that the margin of error is 3 percent, she is expressing confidence that the randomness of the universe can only mess up the answer by three percentage points up or down—that the real answer is somewhere between 61 percent and 67 percent. However, this confidence isn't absolute. Randomness is, well, random, and sometimes a quirky and unlikely set of events can throw off the result of the poll by more than 3 percent. However, something so bizarre can occur only fairly rarely; only one in twenty polls like this can suffer from a strange event that messes up the result by more than 3 percent. Most of the time—in

---

* More precisely, statistical error shrinks in proportion to the square root of the size of the sample.

† If you are mathematically inclined, read appendix A for a fuller explanation of statistical error and margin of error.

nineteen out of twenty polls like this—the randomness of the universe screws up the poll's answer by no more than 3 percent.

Still with me? If you don't understand it fully, don't worry. The margin of error is a really hard concept to wrap your head around, and many journalists, even those who regularly report on polls, don't get it. There are two important things to remember about the margin of error. First, the margin of error reflects the imprecision in a poll caused by statistical error—it is an unavoidable consequence of the randomness of nature. Second, the margin of error is a function of the size of the sample—the bigger the sample, the smaller the margin of error. In fact, the margin of error can be considered pretty much as nothing more than an expression of how big the sample is.*

The margin of error is a direct result of the mathematical laws of probability and randomness. It describes a fundamental limitation to the precision of a poll, an unavoidable statistical error that faces pollsters when they use a sample of people to intuit the beliefs of an entire population. There's no way to get around it; the moment a pollster makes a leap of faith and assumes that a sample of people has the same predilections as the entire population, she introduces this error.

The margin of error is a subtle concept, and it's abused and misused by almost everybody who gets their hands on it—especially journalists. When the press uses a poll's margin of error as an all-purpose gauge to declare whether the poll reflects external reality

* If a poll has a sample size of n, a really good estimate of the margin of error is simply $\frac{0.98}{\sqrt{n}}$. This is why so many polls nowadays happen to have a 3.1 percent margin of error: it corresponds to a nice round sample size of 1,000 people.

or not, they are straying from the realm of reality. And when they treat a poll with a very small margin of error as an oracle almost guaranteed to be accurate, they are making a dangerous mistake.

The margin of error only represents statistical error, the inaccuracy inherent to using a sample of a population to try to represent the whole. While that error is extremely important—it cannot be ignored—there are plenty of other errors that creep into polls that aren't reflected in the margin of error. When polls go spectacularly wrong, the problem is almost never caused by statistical error. A more insidious kind of error—*systematic error*—is almost always to blame. However, systematic errors are never included in a poll's margin of error. When journalists use the margin of error as a litmus test to figure out whether or not to believe a poll, they are completely blind to the sources of error that are most likely to render their poll meaningless. Every time a journalist cites the margin of error as a reason to believe the results of a poll, he's doing the logical equivalent of looking only one way before crossing a two-way street. Sooner rather than later, he'll be clobbered by a bus.

Indeed, the history of polling is filled with spectacular accidents—and it's littered with journalistic roadkill.

By rights, it should have been one of the most accurate polls ever produced. In August 1936, the incumbent president, Democrat Franklin Delano Roosevelt, was facing a stiff challenge from Alf Landon, the Republican governor of Kansas. *Literary Digest* magazine began an unprecedented campaign to predict who would win the upcoming election. For two decades, the *Digest* had built a towering reputation for accurate polls—it was the place where political junkies could get their survey fix. Even the magazine's advertisements were couched in the language of polls. ("Leader in every taste test, winner of every digest poll, Heinz aristocrat tomato juice is overwhelmingly elected by flavor connoisseurs everywhere!")

According to the editors of the magazine, the secret to their success—their polls' incredible accuracy—was their enormous sample of voters. After all, conventional wisdom was that the bigger the sample, the more accurate the poll, and each year the *Digest*'s sample got larger and larger. By 1936, the magazine sampled nearly one-quarter of the eligible voters in the United States. Ten million people would receive a ballot in the mail, which they could then return to be tallied by the *Digest.* Merely getting those ballots to their intended recipients was a truly monumental task:

> This week, 500 pens scratched out more than a quarter of a million addresses a day. Every day, in a great room high above motor-ribboned Fourth Avenue, in New York, 400 workers deftly slid a million pieces of printed matter—enough to pave forty city blocks—into the addressed envelops [*sic*]. . . . Next week, the first answers from those ten million will begin the incoming tide of marked ballots, to be *triple-checked*, verified, *five times* cross classified and totaled. When the last figure has been totted and checked, if past experience is a criterion, the country will know *to within a fraction of 1 per cent.* the actual popular vote of forty millions.

Such a vast undertaking cost a "king's ransom," as the editors of the *Digest* pompously put it. "But *The Digest* believes that it is rendering a great public service—and when such a service can be rendered, no price is too high."*

From the point of view of the *Literary Digest*'s editors, the enormous sample size more than justified the cost. If voters returned even a small proportion of the ballots, the sample would still be so large that it would reduce the poll's margin of error to almost nothing—*to within a fraction of 1 per cent*, as the editors made so plain with their italics.

The typical modern poll of 1,000 people has a margin of error of about 3.1 percent. Increase the sample size, and the margin of error shrinks. A poll of 4,000 people has a margin of error of about 1.6 percent; in a sample of 16,000 the margin of error drops again to about 0.78 percent.

Week by week, the results trickled in as people returned their ballot cards; each week the *Literary Digest* clucked and strutted, blaring the accuracy of its polls and the importance of its huge sample. Right out of the starting gate, the magazine had more than 24,000 responses. This would correspond to a margin of error of roughly 0.6 percent—way, way lower than what even most modern polls can claim. By the week before the election, the sample had risen to an unbelievable 2,376,523 ballots, all tabulated by hand. This would correspond to a margin of error of six-hundredths of a percent: 0.06 percent. Though the editors of the *Literary Digest*

---

* News organizations love to pretend that their actions are motivated by a selfless desire to serve the public good. However, as businesses, they're more often than not prompted by corporate self-interest. Several years earlier, the *Digest*'s editor admitted to Congress that the polls, as expensive as they were, were "a business proposition. Attached to each postcard sent for balloting is a subscription blank for the magazine. The returns in subscriptions have been enormous and they have paid the expenses of the polls."

didn't express it in these terms—the term "margin of error" hadn't yet come into fashion—the tremendous sample size gave them enormous confidence in the poll's results. The week before the election, they couldn't resist bragging:

> The Poll represents the most extensive straw ballot in the field—the most experienced in view of its twenty-five years of perfecting—the most unbiased in view of its prestige—a Poll that has always previously been correct.

That perfect record was about to become roadkill: the poll predicted that Alf Landon would beat Franklin Roosevelt by a large margin. The election, of course, was a landslide in the opposite direction.

The *Digest* predicted that Landon would win about 54 percent of the popular vote, and the margin of error was a mere 0.06 percent. Instead, Landon only got 37 percent. It was a huge mistake, more than 250 times larger than the margin of error would seem to allow. At one blow, the *Literary Digest*'s reputation for accuracy lay bleeding at the side of the road; the large-samples-equals-accurate-polls dictum was dead. What went wrong?

To modern pollsters, the answer is not hard to find. It's implied by the *Digest*'s own description of how they picked the ten million voters to survey: "The mailing list is drawn from every telephone book in the United States, from the rosters of clubs and associations, from city directories, lists of registered voters, classified mail-order and occupational data."

The year 1936 was in the midst of the Great Depression. There was a great division between rich and poor; the rich tended to vote Republican and the poor tended to vote Democrat. Phones were still not in the majority of households; those that had phones tended to be richer—and more Republican—than those that didn't. Therefore, using the telephone directory as a way of generating a mailing list introduces a bias because it is a list enriched with Republicans at the expense of Democrats. The same is true of clubs and associations, especially automobile associations. These people leaned Republican too. Occupational data excludes the unemployed, who tend to vote Democrat. So by drawing their names from phone books, lists of club members, and occupational data, the editors of the *Literary Digest* were inadvertently reaching more Republicans than Democrats. Their sample was not truly representative of the voting population of the United States, creating a systematic error in their poll.

Unlike a statistical error, a systematic error like this doesn't diminish as the sample size grows. It doesn't matter whether your sample is a hundred or a thousand or ten million people: the error caused by a poor choice of sample stays large even as the margin of error shrinks to insignificance. The failure to recognize this source of error made the editors of the *Literary Digest* guilty of (ignorant) disestimation: they thought their sample size made their poll accurate to within a tiny fraction of a percent, when in fact it couldn't be trusted within ten or fifteen points.

Even before the fiasco, more sophisticated pollsters recognized that something was wrong with the *Literary Digest*'s methods. In July 1936, a few weeks before the *Literary Digest* poll got under way, George Gallup, the father of modern polling, predicted that the bias in the *Digest*'s sample would lead to a wrong answer. Naturally, the *Digest*'s editor flipped out, exclaiming, "Never before has anyone foretold what our poll was going to show even before it started!" Gallup was right, though. As he foretold, the systematic error distorted the poll's result so much that it erroneously predicted a landslide for Landon.

There was another more subtle form of bias at work that gently made the *Literary Digest*'s sample Republican-heavy. The *Digest* was justly proud of its impressive sample of 2.3 million ballots, but it's important to keep in mind that the magazine had sent out ten million envelopes. This means that the vast majority of voters—more than three in four—tossed their envelope in the trash rather than sending in a response. The *Literary Digest*'s poll counted only the responses from recipients who cared enough to take the time to fill out the ballot and put it back in the mailbox.

People are relatively silent when they're reasonably content, but if they're angry they tend to shout it from a mountaintop. We don't usually fill out a customer satisfaction survey when we're happy with a restaurant's service—but if the waiter was surly and the food was cold, we immediately start looking for a pencil so we can fill in the response card. When surveys and polls depend on a voluntary response, it's almost always the case that people with strong opinions tend to respond much more often than those who don't have strong opinions. This introduces a bias; the poll disproportionately reflects extreme opinions at the expense of moderate ones.

In the 1936 election, the Republicans were out of power. Discontented with Roosevelt's policies and frustrated at their inability to change them, the typical Republican voter was unhappy with the government. A vote for Alf Landon was a vote for change, a vote of dissent. Democrats, on the other hand, tended to be content with the government. A vote for Roosevelt was a vote for staying the course, for the status quo. When the *Literary Digest*'s envelopes arrived in the mail, the angry, discontented Landon voters were much more likely to send in a response than the content Roosevelt ones. As a result, the flood of cards coming back were disproportionately Republican.

Both of the *Literary Digest*'s mistakes had the same effect. Mailing to mostly well-to-do households biased the sample toward Landon voters, as did the "volunteer bias" caused by the reliance upon people's willingness to fill out a ballot and mail it back. With this particular poll, these two errors were enormous, throwing the result off by tens of percentage points, even though the margin of error was incredibly tiny. What doomed the *Literary Digest* poll was systematic error, not statistical error.

Systematic errors are much more dangerous than statistical errors. They can be extremely subtle—they often manage to be completely invisible until they smack you upside the head. They can be difficult (and expensive) to avoid. And they come in a variety of forms, each one deadly in its own way.

George Gallup made his name, in part, by predicting the *Literary Digest* failure. His polls were sophisticated: by "scientifically" choosing samples carefully to make sure that they were representative of the population, Gallup produced some of the most accurate polls of his day, even though he used much smaller sample sizes than some of his more money-flush peers. He gained a reputation for reliability—and he helped renew journalists' faith in polling even before the embers of the *Literary Digest* conflagration had cooled. But even Gallup's "scientific" methods were subject to systematic errors.

The "Dewey Defeats Truman" headline was the consequence of Gallup's most famous failure. In the weeks before the 1948 election, his polls showed that Dewey was way ahead—so much so that his employees stopped conducting polls (which are rather expensive, after all). But public opinion changes over time: in 1948, the support for Truman got stronger in the weeks before the election as people abandoned third parties in favor of candidates who had a chance of winning—and those voters tended to favor the incumbent. Thus sampling too early introduced a bias against Truman. Another subtle error arose from a faulty assumption about undecided voters. Gallup thought that the voters who hadn't thrown their lot in with Truman or Dewey would vote in the same manner that decided voters had—that is, since the majority of decided voters preferred Dewey, the majority of undecided voters would feel the same way. This wasn't true at all: undecided voters behave very differently from those who have expressed a strong preference. In this case, more of them finally threw their lot in with Truman than expected. The assumption that undecideds behave like other voters subtly biases a poll in favor of the person in the lead, making results look more significant than they are. Again, a systematic error killed a poll, causing it to yield the wrong result. Gallup too fell into the trap of (ignorant) disestimation. He thought his measuring instrument was much more precise than it actually was, leading him to underestimate the uncertainties that surrounded the numbers he produced.

Surprisingly, the 1948 result didn't cause any permanent damage to the public's confidence in polling. Gallup himself wasn't deeply shaken by the wrong results—he was convinced that he could find the problems and root them out, gradually perfecting the art of polling. He kept faith in his methods, even though systematic errors kept cropping up to destroy his polls. For example, his failure to control for respondents' religion might have been responsible for the wrong prediction that Kennedy would win the 1960 election by four points. (In fact, it was a squeaker where a mere 0.25 percent of the popular vote separated the candidates.)

Systematic errors are bewildering in their variety, complexity, and subtlety. Some never get fully explained. In the early 1990s, pollsters in the United Kingdom never were able to figure out why polls consistently underestimated the votes for the Conservative Party. Without coming up with a real explanation, they dubbed the error the "Shy Tory" effect. It disappeared after a few years, its departure as mysterious as its arrival. Some systematic errors are downright creepy. For example, a careful study of certain kinds of polls demonstrates, beyond a shadow of a doubt, that people regularly lie to pollsters, making their data all but meaningless. When this happens, the polls stop reflecting reality at all. They become Potemkin numbers.

Every few years, someone tries to figure out our society's sexual habits with a poll or survey. Some of these surveys in particular are extremely sophisticated and professional—they use expert statisticians and are backed by government money. But every single one returns utter garbage. It's because we humans are liars—we simply can't help it.

In 2007, to much fanfare, the Centers for Disease Control, the preeminent organization for generating health statistics in the United States, released a report about Americans' sex lives. The element that got the most attention in the press was, naturally, the most salacious: how many sexual partners the typical American has had. "Average man sleeps with 7 women," blared the Associated Press. That is, the typical American man has sex with seven women in his lifetime. That statistic in itself was not so surprising. But it becomes shocking when you combine it with another statistic from the same study: the typical woman sleeps with four men in her lifetime. This is impossible.*

Every time a man has sex with a woman, a woman has sex with a man. Each act of heterosexual intercourse increases the average male's number of sexual partners; however, it must increase the average female's number of sexual partners by the same amount.† The male average *must* equal the female average; there's absolutely no way that men and women can have different average numbers of sexual partners. The survey was simply wrong. Its finding that men had seven partners and women had four is just not plausible. It's a fiction, a Potemkin number.

What went wrong? It's simple: somebody's lying. Either the men are exaggerating their exploits, or the women are downplaying theirs—or both. There's no doubt. Even in a clinical environment, even when the pollsters ensure complete confidentiality, even when there's absolutely nothing to gain by lying, people always lie about the number of sexual partners they've had. It happens over and over again. In France, in England . . . they tell fibs about their sexual history.‡ No matter how the poll is run, you're guaranteed to find out that the data are wrong, because people are habitual liars.

---

* It is impossible if we're talking about the mean number of sexual partners, as implied by the Associated Press's use of the word "average." However, the study was actually referring to the *median* number of sexual partners, not the mean. Technically, it's *possible* to have the median number of male partners differing dramatically from the median number of female partners, but it's highly, highly unlikely—and in fact polls that report the mean number of sexual partners also have the men-have-more-partners-than-women flaw.

† This assumes that the number of males is roughly the same as the number of females, which of course is true.

‡ The evidence of lying is clear. In 2003, a study showed that if you hook people up to a fake lie detector machine, the gap between the men's and women's answers narrowed dramatically.

People lie because, on some level, they want to be liked and respected by everyone—including the person who's asking the questions in a poll. Even if they're never going to see the person asking the questions again, subjects tends to give answers that (they think) put them in a favorable light. It's not just on matters sexual. People try to project the noblest image they possibly can to a pollster, whether or not that image is grounded in reality.

For example, less than a week after Hurricane Katrina devastated New Orleans, the Associated Press, teaming up with the polling company Ipsos, sought a variety of opinions from Americans about the disaster. One interesting tidbit: more than two-thirds of Americans claimed to have already donated money to Katrina relief efforts. Conservatively, this would mean that about sixty million households had donated money—all within a week of the disaster. This, naturally, would mean that an enormous amount of money would quickly flow to the disaster victims.

How much money? It's hard to say for sure. The Salvation Army reported at the time that its average Katrina-relief donation was around two hundred dollars. That's probably the typical size of a Katrina donation, but to be on the safe side, let's assume that other relief organizations didn't do as well, and that the average donation was much smaller than what the Salvation Army saw—say, fifty dollars. So if we have sixty million households, each donating fifty dollars, that would mean a total donation of a whopping three billion dollars.

Americans can definitely be generous in times of crisis, but they weren't the selfless philanthropists that the poll would imply. At the time the poll was conducted, the total amount of money raised for Katrina victims was about six hundred million dollars—around one-fifth of what you'd be led to expect from the poll numbers. If you take into account the fact that much of that six hundred million came from corporations and not from individuals, the mismatch is even worse. The conclusion is obvious: there's no way that two-thirds of Americans had truly donated money in the week after the disaster. It's a Potemkin number. The truth is probably less than half of what the poll reported. Of course, this means that the majority of the people who said that they had donated were probably lying. People lying is a systematic error, not a statistical one, so the 3 percent margin of error (dutifully reported by the news media) was more or less irrelevant to the accuracy of the poll.

Sometimes the reasons for lying go beyond simple self-aggrandizement. Sometimes there's outright malignant deceit. People occasionally attempt to game a poll for their own benefit. Early in the 1996 Republican presidential primaries, exit polls—polls performed upon voters just as they leave the site where they cast their vote—showed that long-shot far-right candidate Pat Buchanan had beaten the more mainstream Bob Dole in Arizona. Pundits on the networks immediately started chatting about how Buchanan was a credible challenge to Dole. But when the returns—the actual count of ballots cast—came in, Dole had beaten Buchanan by more than two percentage points. Why? According to CNN's political director, it was because Buchanan voters "seek out the exit poll takers" in order to inflate Buchanan's showing in the polls.* Manipulating an exit poll is not that hard to do. If you linger near a polling place or walk toward a pollster (who can usually be identified by the clip-

* When the networks discovered the error, executives said that the problem would likely make them more careful when it comes to using polls to project the winner of a race. Every few years, like clockwork, an exit poll fails in an embarrassing way. Like clockwork, the media promise to be more careful next time. And nothing ever changes.

board he's carrying), you can make it more likely that you get included in the sample, unless the pollster is really paying attention. This overrepresentation can make it look as if your candidate poses a more serious threat than he actually does, perhaps garnering him some more followers.

It's not just Americans who lie to exit pollsters. In 2006, political observers were stunned when the Palestinian elections ended in a tremendous upset. Polls—including exit polls—had predicted that the Fatah party would beat its rival, Hamas, by about six percentage points. Instead, Hamas won by a few points. Again, the mistake dwarfed the polls' margins of error. In fact, the reason for the mistake was that Hamas supporters pulled an anti-Buchanan. They tried to make Hamas appear less popular than it really was, by either lying to pollsters or refusing to talk to them. Why? One of the poll organizers told the *Wall Street Journal* that "Hamas sought to influence the outcome of the exit polls in the hope that the exit poll results would not alarm Fatah armed groups for fear [Fatah] would seek to harm the electoral process (for example, by burning ballot boxes as happened in previous elections at the local level)." If the government burns your ballot boxes when there's too much support for your party, it makes perfect sense to lie in a poll. Unfortunately, poll manipulation is all too common. There are plenty of other actors who try to manipulate the outcome of polls. The worst are the pollsters themselves.

Since the dawn of polling, researchers have known that the outcome of a poll can be affected quite dramatically by the wording of the questions in the survey. The choice of words in even the most neutral-sounding phrase sometimes subtly steers a subject to answer in a particular way. For example, in the early 1990s, Gallup (teaming up with CNN and *USA Today*) asked Americans whether they supported bombing Serbian forces in Bosnia. The response was extremely negative: 55 percent against to 35 percent in favor. The very same day, ABC News asked a similar question in a different poll, and the answer this time was overwhelmingly positive: 65 percent supported airstrikes, while 32 percent opposed them. How could two polls give such amazingly different results? The mismatch was because of a subtle difference in the way the questions were worded. In the Gallup poll, the United States was doing the airstrikes. In the ABC poll, the bombings were carried out by the "United States, along with its allies in Europe." Such a tiny change in wording colored the questions, and Americans reacted differently based on whether they were given the impression that the United States was acting unilaterally or not. This is an enormous source of systematic error—an infinitesimal shift in phrasing can completely change how people answer a poll question. Even something as seemingly insignificant as the order of questions can make a tremendous difference.

When there's a tricky issue likely to cause problems, some of the better-funded polls ask two or more sets of people one of two or more different versions of the same question to try to ensure that the wording isn't messing with the results.* Unfortunately, this makes the poll more expensive and increases the margin of error,

* For example, in January 2006 the *New York Times* asked people two questions about the controversial warrantless wiretapping program that was then coming to light. One question had a clause about Bush's claim that the wiretaps were necessary to fight terrorism. The public came out in favor, 53 percent to 46 percent. An alternate wording of the question omitted Bush's claim, and the public came out against, 50 percent to 46 percent. Few other news organizations were as careful. For example, another major polling firm, Rasmussen, asked a question that completely failed to mention anything about not needing a warrant or a court order for the wiretaps. Not surprisingly, its survey found that the public was decidedly in favor of wiretaps.

since splitting a question into two different variants means that each variant gets answered by half the number of subjects. (Worst of all from the perspective of the polling organization, if two phrasings get vastly different answers, it might reveal that the poll is meaningless.) Thus most polls don't control for systematic errors caused by wording, even when it's desperately needed.

And it is desperately needed quite often. Anytime there's a politically sensitive or awkward issue, the pollster's choice of language is going to affect the outcome. The term "pro-life" evokes very different reactions than the term "anti-abortion." People will be less concerned about "collateral damage" in a war than they will be about "dead civilians." The same respondents who oppose "torture" will be more accepting of "enhanced interrogation techniques." Some of these words and phrases have such strong connotations (it's really tough to admit that you are a fan of torture) that it's almost impossible to cut through the fog of words to get at people's true feelings. Even innocent-seeming choices can be loaded: "President Obama" is more deferential than "Barack Obama," so the former will tend to poll better than the latter. Any choice of phrasing will be biased in some way, creating a large source of error.*

From a mathematical point of view, this dependence on question phrasing is a systematic error that prevents polls from reflecting reality. To some pollsters, though, reality is merely a burden. They can make more money if they give their clients what they want—and if their clients want a Potemkin number, they seem to be willing to provide it.

---

* A fun example: a poll where you ask half your subjects, "Do you think it's acceptable to smoke while praying?" and ask the other half, "Do you think it's okay to pray while smoking?" Even though the questions are asking about essentially the same behavior, the polls would have very different outcomes.

In April 2005, a deeply divisive drama was drawing to a close in the United States. Terri Schiavo was a woman who had been in a persistent vegetative state for more than a decade. Her husband, Michael, wanted her feeding tube removed; her parents wanted Terri to be kept alive. After a long battle in the courts and a great deal of posturing by politicians, Michael prevailed and the tube was removed. It was a grisly business; unable to feed herself, Terri was slowly starving. Though people felt strongly on both sides of the issue, the polls in March seemed to show that Americans tended to favor the removal of the tube. An ABC poll, for example, had 63 percent in favor of the feeding tube's being removed.

The Christian Defense Coalition, horrified by the results, evidently sought to bend public opinion to reflect their reality. So they hired Zogby International, one of the premier polling firms in the nation, to conduct a new poll. Zogby agreed, and when the results were released in early April, an article reproduced on Zogby's website trumpeted the results: "A new Zogby poll with fairer questions shows the nation clearly supporting Terri and her parents and wanting to protect the lives of other disabled patients." What were these fairer questions exactly? A small selection:

- Do you agree or disagree that the representative branch of governments should intervene when the judicial branch appears to deny basic rights to the disabled?
- Do you agree or disagree that the representative branch of governments should intervene when the judicial branch appears to deny basic rights to minorities?

- Do you agree or disagree that it is proper for the federal government to intervene when basic civil rights are being denied?

- If a disabled person is not terminally ill, not in a coma, and not being kept alive on life support, and they have no written directive, should or should they not be denied food and water?

- Michael Schiavo has had a girlfriend for 10 years and has two children with her. Considering this, do you agree or disagree that Michael Schiavo should turn guardianship of Terri over to her parents?

This wasn't an attempt to discover public opinion. It was an attempt to shape it. The questions about minorities and the disabled were intended to frame the issue as one where Schiavo's civil liberties were being violated. Then the use of "denied"—a very evocative word—in the key question makes it hard for anyone but a monster to say that Schiavo (however abstractly she's presented) should be disconnected from the feeding tube. And then, in case you had any doubts about where your loyalties lie, the poll paints Michael Schiavo as a philanderer. Naturally, the Zogby poll had quite different results from the other polls that used less inflammatory language—results that the Christian Defense Coalition clearly wanted. Another client pleased, reality be damned.

Polling organizations make money by producing Potemkin numbers for their clients. Antievolutionists, environmentalists, interest groups of all sorts—all line up for the privilege of telling Americans precisely what they think. No cause is too silly. Even alien-abduction advocates have paid for polls to prove what they know in their guts to be true. In 1992, the Roper Corporation ran a large poll that was designed to demonstrate that some Americans (2 percent, it turns out) had been abducted by aliens. It's a lucrative business to engineer data to support falsehoods.

Journalists, at least the more sophisticated ones, are well aware of the huge problems with polling, yet they're willing to ignore them. Even the dimmest of pollsters knows that Internet polls are utterly worthless. Ignoring the fact that Internet viewers aren't representative of the general population, ignoring the volunteer bias caused by low participation in online polls, Internet polls are still so flawed that they have been considered the height of polling silliness for more than a decade. The major reason is that they can be easily manipulated with the slightest of effort, and often are. Internet polling is why "Hank, the Angry Drunken Dwarf" beat out Brad Pitt as People.com's "Most Beautiful Person of the Year" in 1998. It's also why Stephen Colbert, modern master of Internet survey manipulation, won polls to name a module of the International Space Station as well as a bridge in Hungary after him. Internet polls have no basis in reality whatsoever. Yet CNN.com has an Internet poll on its front page every day. It's not for information; it's for titillation.

Reporters are supposed to hold the truth sacred above all else, yet they don't seem fazed at all when reporting polls that are utter tripe—so long as they have sufficiently small margins of error. They aren't even bothered by the most glaring contradictions. At the very end of 2006, the Associated Press and AOL teamed up to conduct a poll designed to gauge Americans' feelings about the coming year. As soon as the pollsters finished gathering the data, the Associated Press wrote not one but two stories about their poll.

The first: "AP Poll: Americans optimistic for 2007."

The second: "Poll: Americans see doom, gloom for 2007."

That's right. Both stories were about exactly the same poll. Both stories were published by the same organization—the one that performed the poll. Both reporters consulted AP's manager of news surveys, Trevor Tompson, before writing their articles. Yet the two stories used the same data to come to exactly the opposite conclusions. Nobody at the Associated Press seemed to worry that they were publishing nonsense; nobody seemed concerned in the slightest that reporters were presenting what I believe are meaningless, self-contradictory Potemkin numbers in the guise of objective truth.

In my opinion, there's a legal term that describes this sort of behavior: *absolute malice*. This term makes journalists shudder, because it represents a failure so great that it pierces a hole in the First Amendment shield that protects the press. A reporter or news organization acts with absolute malice either when it publishes something that it knows is false, or when it acts with reckless disregard for the truth. When it comes to polls, I believe that the news media as a whole—not just the Associated Press—regularly act with reckless disregard for the truth.

Sure, news organizations try to dress up polls in the guise of objective reality; they publish a margin of error to tell the public how far the poll can be trusted. But at best, that margin of error leads to the sin of disestimation; since it doesn't take into account systematic errors, it deliberately overstates the accuracy of the poll. At worst, it gives journalists license to ignore even the most obvious problems that plague a poll—it's a tool they use to dress up nonsense in the guise of a fact. Journalists know that Internet polls are worthless, yet such polls regularly make the news. They insist on using telephone polls even though they know that such surveys are inherently biased. (After all, they reach only those rare people who don't slam down the phone when they get a call from a random stranger asking them to answer a few questions.) Their faith in polling is never shaken, even when a high-profile poll comes back with the wrong answer and embarrasses them.

In the early evening of November 7, 2000, exit polls indicated that Al Gore had won Florida. All the news networks rushed to make the announcement of Gore's victory. Of course, the announcement was premature; the vote was far too close to call. Within a few hours, the news media retracted their announcement with much hand-wringing and self-effacement. "If you're disgusted with us, frankly, I don't blame you," intoned CBS news anchor Dan Rather. Chastened and humbled, the news anchors vowed never again to make such an idiotic error; they would be more careful with their exit polls. The resolve lasted just a few hours—until the moment that brand-new exit polls indicated that George W. Bush had won the election. Within moments, Gunga Dan was again coronating the new president. "Sip it, savor it, cup it, photostat it, underline it in red, press it in a book, put it in an album, hang it on the wall," he declared. "George Bush is the next president of the United States." Again, it was a false alarm. The race was still far too close to call. Again, the networks sheepishly retracted their mistake—but not before Gore made a phone call conceding the race, which he too had to rescind.

Polls are like a drug. The news media can't swear off them for even a few hours, no matter what damage they do to their reporters' reputations. The networks spend enormous amounts of money supporting their habit, even dropping tens of millions for a fix when they're really jonesing, such as on election night.

Election night is the most dramatic manifestation of the news media's addiction to polls. Whenever there's a national election, the networks set up a massive and vastly expensive exit polling system. Yet all that their money and effort buys (at best) is a few hours—the audience gets unofficial word about who won an election at 10 p.m. rather than waiting until 7 a.m. for the official returns. Ask yourself: Would your life be diminished at all if there were no exit polls? Would your life be worse off if there were no polls at all? Unless you're a marketer or politician, the answer is probably no.

Yet when it comes to polling, all journalistic objectivity and skepticism seem to go out the window. The pull of the poll is so great, the addiction to pseudoevents is so desperate, that the media forsake fact. The news media don't seem to care if a poll represents any form of truth. It's proofiness, pure and simple.

Sadly, even truth itself isn't immune from the ravages of proofiness. Reality can be fuzzy; even events whose outcomes should be clear-cut can be cast into doubt by mathematical malfeasance. Even something as seemingly idiot-proof as tallying the number of objects in a pile is regularly messed up by errors inherent to the process of counting. Worse yet, it is distorted by proofiness whenever those objects being counted are valuable. Like votes.

5

Electile Dysfunction

It's not the voting that's democracy, it's the counting.

—Tom Stoppard, *Jumpers*

Minnesota's Senate election officially went off the deep end on December 18, 2008. That was the day that lizard people ate a vote in Beltrami County.

Beltrami County is in the frigid north of the state, up near the Canadian border—not thought to be the best environment for lizard people. However, one voter's ballot implied that there was a small community of lizard people in Minnesota who saw fit to run for public office. Not just one office, but all of them. The voter had scrawled the same name in the write-in space for each race: "Lizard people." U.S. president: Lizard people. U.S. representative: Lizard people. Mayor: Lizard people. State representative? Soil and water conservation supervisor for District 3? School board member? Lizard people, lizard people, lizard people. The voter even was careful to bubble in the oval next to each write-in candidate. Except in one case: the race for U.S. senator.

On election day, Minnesotans knew that the Senate race between Republican incumbent Norm Coleman and his challenger, Al Franken, was going to be close. It had been an ugly campaign, and the candidates had been neck and neck for weeks. However, nobody predicted just how close it would be. The two would be separated by a few thousandths of a percent of the votes cast—just a tiny number of ballots would determine who won the race and who lost. As a result, lawyers from both parties descended on Minnesota to see if they could wring out an extra handful of votes for their candidate—or steal a few from their opponent—by whatever means necessary. Every single vote had to be fought over, as each one might mean the difference between victory and defeat. This is where the lizard people came in.

Our intrepid voter from Beltrami County did indeed write in a vote for lizard people for Senate as he did for the other offices, but there was an important difference. He didn't bubble in the oval next to his write-in vote; instead, he messily filled in the oval next to Al Franken's name. Initially the vote had been counted for Franken; when the ballot had been fed through the scanning machine, the scanner only spotted (and recorded) the Franken vote, ignoring the lizard people vote. But sharp-eyed Coleman partisans had spotted the write-in lizard people vote and argued that the ballot should be thrown out. After all, the voter had seemingly cast two votes—one for Al Franken and one for lizard people—in a race where you could only vote for one candidate. If so, this would invalidate the ballot; Franken would lose a vote. So should Franken keep his vote, or should he lose it? This was the question before a five-person canvassing board—four Minnesota judges along with the secretary of state—tasked with counting votes and certifying the winner of the election.

Of all the questions that came before the canvassing board, that of the lizard people was perhaps the most absurd. Even though it would seem to be beneath the dignity of the judges to argue about lizard people, there was in fact a legal issue at stake. If there truly was a legitimate vote cast for the lizard people in addition to the one for Al Franken, it should be counted as an overvote. If the vote wasn't legitimate, it should go for Franken. So the question before the judges was: was the vote for lizard people legitimate?

Oddly enough, it didn't matter that the oval next to the write-in candidate spot wasn't bubbled in. Minnesota law is clear on that point: "If a voter has written the name of an individual in the proper place on a general or special election ballot a vote shall be counted for that individual whether or not the voter makes a mark . . . opposite the blank." Thus the write-in vote for lizard people was valid, meaning the voter had overvoted and the ballot should be discarded. But nothing's ever quite that simple.

Marc Elias, Franken's perpetually sweaty lawyer, came up with a brilliant last-ditch argument to save the vote for Franken: he asserted that "lizard people" wasn't a real person. Since the voter didn't write the name of an actual *individual* in the write-in spot, the write-in vote shouldn't be counted as genuine. That's when the discussion got downright surreal. Coleman lawyers countered by insisting that "Lizard People" was a real person. All sense of decorum disappeared as the judges on the panel promptly started fighting among themselves.

"If it said 'Moon Unit Zappa,' would you say, 'Oh, no, there is no such person as Moon Unit Zappa'?" asked judge Eric Magnuson. To drive the point home, he continued emphatically, "You don't know that there isn't someone named 'Lizard People.' You don't. You and I don't."

Asked Judge Edward Cleary, "Isn't 'people' plural? I mean, how do you have an individual named 'people'?"

"Well," countered Magnuson, "you could have a last name Sims." Soon, Judge Kathleen Gearin chimed in: "I have to admit that I've had someone appear in front of me in court whose name was 'People,' 'Peoples,' spelled like that—"

"First name 'Lizard'?" interrupted Cleary, incredulously.

Yes, indeed. A room full of hot-shot lawyers and eminent judges was engaged in a serious debate about whether Mr. Lizard People was a real human being. It was a hard-fought battle because the Minnesota election was so close that the answer might mean the difference between victory and defeat.

Even though extremely close elections bring out the most bizarre and irrational behavior that human beings are capable of, the lizard people question took even the most jaded election observers by surprise. It quickly became a symbol of how screwed up—how detached from objective reality—Minnesota's election had become.

Most elections are reasonably rational and genteel affairs, at least once all the votes have been cast. Sure, attack ads might pollute the airwaves for months before an election, and candidates will hurl nasty rhetoric at each other, but after the ballots are tallied, everybody agrees about one crucial fact: who won the contest. The loser gives a heartfelt concession speech full of hope mildly tempered with regret, perhaps sheds a few tears, and then congratulates the victor. No matter how vicious the fight, the battle ends as soon as the polls close. When the votes are finally tallied, everybody—even the loser—reaches an accord about the outcome. Most of the time.

When an election is extremely close, counting the ballots might not end the battle. That's when lizard people begin to crawl out of the woodwork. Within hours of the polls' closing, both sides accuse the other of trying to steal the election. The candidates and their lawyers maneuver for advantage, trying to find arguments that will secure victory even when they're lacking a resounding mandate from the voters. Both sides attempt to weasel their way into office. This is the perfect environment for proofiness.

Indeed, Minnesota's 2008 Senate election was ground zero for electoral proofiness, just as the 2000 presidential election in Florida was nearly a decade earlier. Minnesota, like Florida, was soon deluged with twisted arguments, distortions, and outright lies as both parties jockeyed for position.

In truth, there was a mathematically and legally correct way to settle the Minnesota and Florida elections fairly; unfortunately, the proper procedure was rather disturbing. The correct method for determining the victor of these elections happened to be the only solution that no candidate would ever have been able to embrace. As a result, the outcome of the Minnesota election, like the outcome of the 2000 presidential election before it, had absolutely nothing at all to do with logic or mathematical truth. These elections were determined by nothing more than proofiness.

It shouldn't be this way. Elections should be foolproof by now. After all, we humans have been voting for more than two millennia. Ancient Athens, for example, regularly held plebiscites of all kinds, and their voting procedure was as simple as can be. Citizens would write names on potsherds and dump them in the marketplace. Magistrates would then count the potsherds, one by one, reading aloud the names inscribed upon them. Two and a half millennia later, voting is almost exactly the same as it was in ancient Athens. True, the technology has changed somewhat over the centuries. Instead of voting by writing a name on a piece of pottery, we do it by writing on a piece of paper or punching a hole in a card or flipping a lever in a voting booth or bubbling in a circle on a Scantron sheet. But fundamentally, the ballot is the same—it's just in a slightly different form.* Just as in Athens, the election is decided when humans get together (ostensibly) under the eyes of the public and enumerate the votes, one by one. Determining the winner of a vote is as simple as the act of counting; it's as easy as one, two, three.

After twenty-five hundred years, you'd think that we would have figured out how to get this simple little procedure right. Yet even under the best of circumstances, we manage to mess it up. The incredibly forthright act of counting one, two, three regularly falls victim to proofiness.

At first glance, Minnesota would seem to be the most unlikely spot in the nation for the site of a bitter and dirty electoral battle. The citizens of the state have a reputation for being ridiculously earnest and forthright. In other places, election wrongdoing might involve stuffing ballot boxes, phony voter registration, or suppressing votes. In Minnesota, the biggest electoral scandal in recent history had to do with Twinkies.

Make no mistake, though. Twinkiegate was big news for Minnesota; it even made national headlines. In 1986, a grand jury indicted George Belair for maliciously distributing $34.13 worth of

---

* Electronic voting *is* a little different, in that there need not be a physical token to count. More on this in appendix B.

Twinkies, Ho-Hos, and doughnuts to senior citizens while he was running for the Minneapolis city council. "How could anyone bribe someone with Twinkies?" Belair asked, baffled, as he was hauled off to jail. But Belair had in fact fallen afoul of a state law that forbids candidates from providing food or entertainment to voters. The case ended a few weeks later when the charges were dropped by a judge.

What's more, Minnesota had about the most modern and well-thought-out election laws on the books. After the *Bush v. Gore* fiasco, a number of states, including Minnesota, reworked their laws and regulations to try to avoid another electoral disaster. Minnesota's laws were extremely precise—the rules about how to interpret questionable ballots were, theoretically, all decided ahead of time. (No hanging-chad debates for Minnesota!) The state's statutes were so prescriptive that they even dictated how to stack ballots for counting—in groups of twenty-five, placed crosswise. Perhaps the most reassuring element of Minnesota's election laws was that they created a built-in error-checking mechanism to ensure that ballots were being counted properly. No matter how close an election is, a selection of precincts all around the state have to recount all their ballots by hand to ensure that the counts are accurate. Further, if the vote is extremely close—if the margin between the candidates is less than half a percent of the number of votes cast—it automatically triggers a statewide recount.

Thanks to the modern electoral laws, the uniform voting procedures, and the honest character of the citizens of the state, it's hard to imagine better conditions for a fair and uncontroversial election than Minnesota in 2008. Even though the election was looking fairly tight by the end of October—polls, for what they're worth, disagreed about who was in the lead—nobody anticipated the nastiness, vitriol, and proofiness that would envelop the state for months to come.

True, the campaign itself had been ugly, especially by Minnesota standards. The challenger, Al Franken, an ex-comedian perhaps best known for his book *Rush Limbaugh Is a Big Fat Idiot*, was an easy target for ridicule. His opponent, incumbent senator Norm Coleman, was no prize package either. Judging from what his detractors said, they seemed less troubled by allegations of Coleman's corruption* than they were repulsed by his bouffant hairdo, unnatural tan, and Chiclet teeth. The battle royal between these two characters turned quite a few voters off; rather than cast a ballot for Franken or Coleman, roughly 15 percent of voters chose to throw their vote to a third-party candidate with no chance of winning the election.

Everybody knew it would be a close race. Even so, the results were astonishing. The first returns, reported on election night, had Norm Coleman ahead by only 725 of the nearly three million votes cast: a margin of only about thirty-thousandths of a percent: 0.03 percent. The Associated Press declared Coleman the victor early in the morning, but quickly rescinded its announcement when news editors realized that the battle was far from over.

The morning after the election, a tired but confident Coleman declared victory: "Yesterday, the voters spoke. We prevailed." He then suggested that Franken short-circuit the automatic recount mandated by Minnesota law by conceding. "If you ask me what I would do, I would step back," he said. "I just think the need for

---

* Coleman was dogged by allegations, which turned into an FBI investigation, about improper gifts given to his wife by a major campaign contributor. However, neither Coleman nor his wife was charged with any wrongdoing.

healing is so important, the possibility of any change of this magnitude in this voting system is so remote, but that would be my judgment." His confidence was unjustified. Coleman's lead was already slipping away.

By the afternoon after the election, the 725-vote margin had dropped to 477 and then to 465. Some changes were attributed to data entry errors. At one precinct, for example, a clerk had accidentally typed "24" votes for Franken instead of "124." (Even if clerks were somehow able to count ballots with perfect accuracy, it means nothing if they can't enter data into a computer correctly.) Absentee ballots were also changing the numbers; some had been included in the initial counts, while others took longer to tally. A week after the election, the margin between the two candidates stabilized at a 206-vote lead for Republican Norm Coleman.

It's hard to fathom just how unbelievably close a contest that number represents. An election where the two candidates are separated by a percentage point is considered very close. If Coleman were ahead of Franken by 1 percent, that would have meant a 28,000-vote lead. To be ahead by only 206 votes means that Coleman had a lead of about seven-thousandths of a percent—less than 1 percent of 1 percent of the total votes cast. This is a mind-bogglingly tiny margin. If we elected people based upon their height instead of the number of votes they received, this election would be equivalent to having one candidate taller than the other by no more than the width of a human hair.

When "close" means 28,000 votes, hundred-vote, twenty-five-vote, or ten-vote errors (which are surprisingly common in an election this big) don't make a real difference. A few hundred votes here or there won't swing the election, so they are ignored—they don't make headlines. We're blissfully unaware that they happen regularly. However, when an election is *really* close—when the margin is a fraction of a percent—those mistakes suddenly make a huge difference. This is why election officials pray for electoral blowouts; big margins cover mistakes, while close ones bring every little glitch into sharp relief.

That is what happened in Minnesota. Because the margin between Franken and Coleman was so small, Minnesota's electoral errors seemed to be huge, even though they were nothing unusual. When those errors were corrected—and when, inevitably, new mistakes were made—the margin between the two candidates changed by a tiny fraction of a percent. With an election this close, even a few hundred votes is a distressingly large amount.

As Minnesota officials began correcting errors, the margin between Coleman and Franken got narrower and narrower. Though their candidate still held the lead, the *Wall Street Journal* cried foul, accusing Democrats of hoping to "[steal] a Senate seat for left-wing joker Al Franken." The beadiest-eyed of Coleman's lawyers, Fritz Knaak, immediately told the press that the changes in vote tallies were "statistically dubious." (When lawyers appeal to statistics, you can be certain that there's proofiness about.)* The reasoning was that correcting errors shouldn't change the vote tally by much—mistakes in favor of one candidate should cancel out mistakes in favor of the other. However, this is wrong. Errors can favor one candidate or another—the errors need not behave like coin flips coming up heads 50 percent of the time and tails 50 percent of the

* There was some old-fashioned political mythmaking too. Knaak accused an election director of mishandling ballots, driving around with them in the back of her car. The accusation was repeated by the *Wall Street Journal* and a number of other news outlets—Sean Hannity harped on it for several nights running. It turned out to be false.

time. Because of this, correcting those errors will not necessarily "cut both ways" as the *Journal* put it.* In this particular election, there were a few reasons why the errors favored Coleman, so correcting them favored Franken.

One reason was that absentee ballots, which are used by a slightly different population than the people who go to the voting booth, tended to favor Al Franken more than the rest of the state's voters. Thus, problems specific to counting absentee ballots (and there were many of them) tended to rob Democrats of their votes more often than robbing Republicans. Another reason is that cities tend to vote Democratic and rural areas tend to vote Republican. Polling places in cities tend to have more voters, more machines, more staff, and more things that can go wrong. And when things do go wrong in big counties, the mistakes tend to be larger. For example, a hiccup in a voting machine in Ramsey County—home of St. Paul—swallowed 171 ballots that had to be added to the total. In Minneapolis, 133 votes in an envelope went missing sometime between the election and the recount. In both of these cases, the bulk of those votes favored Franken, not Coleman.

There were, of course, errors that favored Coleman rather than Franken. During the post-election audit, the hand recount in random precincts required by Minnesota law upped Coleman's lead to 215 votes from 206. And during the painstaking hand recount that followed, many more errors were corrected, some of which were to Coleman's benefit. These errors weren't hard to find—after all, hundreds upon hundreds of votes had vanished.

---

* In fact, even 50/50 errors don't always "cancel out" and can be expected to produce fairly large swings in favor of one party or another on occasion.

Voting has a lot in common with polling. They have a similar purpose; both voting and polling attempt to figure out what the population is thinking about a subject. Also, like polling, voting is subject to error, though the errors are of a different sort. Voting, by its very nature, avoids many of the problems that plague polls.

Polling tries to get at an underlying truth, but that truth is obscured by errors, both statistical and systematic. When pollsters ask a sample of a few hundred or few thousand subjects a question, there's the unavoidable statistical error that is introduced by the leap-of-faith assumption that the sample represents the beliefs of the entire population. The laws of randomness ensure that even under the best of circumstances there's an underlying error that obscures the truth. And circumstances are usually not that great. Pollsters have great difficulty getting a sample that is truly representative of the whole population. Often this causes a systematic error—a bias—that obscures the truth even more. And of course, if a poll is badly worded or improperly conducted, it can cause systematic errors that can make your poll all but meaningless.

Voting avoids these problems almost entirely. In a vote, the truth—the victor of an election—is determined by the will of the population that comes to the voting booths and casts a vote. Unlike a poll, which queries a sample of a population, a vote uses the results from the entire universe of voters who cast a vote. The people whose opinions you are trying to figure out are, by definition, exactly the same ones whose votes you are counting. There is no leap of faith that comes with assuming that a sample truly represents the opinions of the whole population, so there's no statistical error at all. Similarly, you don't have to worry about pulling a *Literary Digest*—since you're looking at the entire universe of votes cast rather than a sample, you don't have any systematic error due to a nonrepresentative sample. The votes that are counted and certified are, by definition, perfectly representative of voters whose opinions need to be considered.* Finally, the issues put to a vote are (generally) so black and white that there isn't (generally) a systematic error caused by the phrasing or presentation of questions. In theory, vote tallies should be fairly "pure" numbers, close to the realm of absolute truth.

Because of this, voting is pretty much immune from the big errors that obscure the truth in polls.† The huge, multiple-percentage-point errors that make most polls meaningless simply aren't germane to elections. But this doesn't mean that elections are error-free. Far from it. It's just that the errors are more subtle. They're smaller—a fraction of a percentage point of the total votes cast—but they're there.

This shouldn't come as a surprise to anyone reading this book. There's no such thing as a completely pure number, no such thing as a measurement that's always perfect. The act of counting votes is a measurement. And like any other measurement, it is error-prone. The act of counting is imprecise, and the degree of imprecision depends upon what you're counting and how you're counting it.

In the 1960s, the navy was getting quite annoyed that they couldn't reliably keep track of the material that they stored in warehouses. Any catalog of material in a warehouse had huge errors,

* Whether the people whose votes are counted really represent the democratic ideal of the "will of the people" is a deeper question that will be left to the next chapter.

† Many journalists don't understand this. During the recount in Minnesota, Minneapolis's ABC affiliate commissioned a poll to see who would win the election if a revote were held. I'd love to meet the genius who thought that a poll with a 4.2 percent margin of error could give any insight into an election where the candidates differed by 0.007 percent of the votes cast.

even after the warehouse staff went around physically inventorying everything in the building. Fed up, they asked an industrial engineer to find out what was going on. The engineer was surprised to discover that when he had warehouse workers count and then recount the inventory, the two counts didn't match up—there was a 7 percent difference between the two inventories. The very act of counting was hugely error-ridden. As one inventory specialist told me, "Humans counting things, I don't care what the things are, they're going to be off."

So long as we count votes, we make errors. And so long as we make errors, we can't be perfectly precise; there's inherent imprecision in our determination of who won a vote. In a blowout election, where the margin between the two candidates is much greater than the errors, the errors can't affect the outcome of the election. But when the election is as close as Minnesota's, the vote might be rendered meaningless—the truth of the real winner of the election will be obscured—if the errors are large enough.

This raises an obvious question: how large are the errors in counting votes? It's hard to say for sure, because there are so many ways things can go wrong. There can be problems with the ballots—they might be read incorrectly by voters or by counters. The ballots might be mishandled and either left out of the tally or double-counted. The ballots might be tallied properly, but people might make stupid mistakes when recording the numbers or adding them together. And then there's the unavoidable error of occasionally losing count—skipping or double-counting a ballot or pile of ballots.

Minnesota's errors were probably quite small compared to those in other modern elections. The ballots all around the state were of roughly the same design, sheets with ovals to be bubbled in, meant to be fed through an optical scanner. These are considered to be the best kind of paper ballot, as they're relatively easy to fill out and easy to interpret. (There are none of the hanging-chad problems that bedevil precincts that use punch cards, for example.) But there were definitely lots and lots of other errors. Hundreds upon hundreds.

It's surprisingly easy to find missing votes if you have the right data. The Minnesota election was commendably transparent. Throughout the count and the recount, the state government shared all of its data with the public and the press. Every day, you could look at the secretary of state's website and download the total votes tallied (and retallied) in each precinct. Shortly after election day, you could also see the voter turnout—how many people showed up at the voting booths on election day and how many cast absentee ballots of various kinds. In theory, these two sets of data should match exactly: in each precinct, for every voter who casts a ballot, you should have a ballot that is counted. But in nearly 25 percent of Minnesota precincts, the two numbers didn't match—there were either more voters than ballots or more ballots than voters. This meant that one of the numbers was wrong.

In some cases, the voter turnout numbers were clearly at fault. Ramsey County's voter turnout lists seemed to have been lovingly maintained by a pack of wild raccoons. One precinct, which had roughly 1,000 voters in the district, somehow claimed to have a voter turnout of 25,000—mostly military overseas ballots. (This error was corrected after several weeks, but it was pretty clear, for various reasons, that none of Ramsey County's original turnout numbers could be trusted.) In other cases, though, the turnout numbers seemed to be solid. This was where it was obvious that people were making errors when they counted votes.

Sometimes a miscount made votes materialize out of thin air. One precinct in Rice County tallied twenty-five more votes than voters who showed up on election day. Fran Windschitl, the election official in charge, suggested that a simple counting error was to blame: "Our accounting procedures dictate that you arrange them in piles of twenty-five; they must have counted one too many piles," he said. Sometimes a miscount caused votes to disappear. In a precinct in Blue Earth County, there were twenty-five fewer ballots than voters who showed up to the polls. Worse yet, a careful look at the data implied that the missing votes had all been cast for Norm Coleman.* (A number of the miscount errors seemed to happen in bundles of twenty-five, which can easily be explained by missing or double-counting one of the crosswise stacks of twenty-five ballots described in Minnesota law.) Sometimes the errors weren't a simple miscount, but had a more complex explanation. All around the state, absentee ballots, particularly those that had been duplicated so they could be fed through the scanning machines, were causing troubles. Sometimes the originals disappeared, and sometimes the duplicates went missing—potentially introducing errors into the final count. All told, there were oodles of miscounted ballots. Hundreds and hundreds of votes had vanished and hundreds of others had materialized. Some of these errors were in favor of Franken, some were in favor of Coleman, and some were a wash.† Many of

* This particular error was (quietly) corrected roughly two weeks after I notified the nonprofit Citizens for Election Integrity Minnesota, who apparently used back channels to notify the Minnesota Secretary of State's office about this and several other problems. Eventually, I wound up writing a four-page letter to the secretary of state and to the two campaigns outlining a large number of precincts that seemed to have made tabulation errors. Quite a few of the inconsistencies I describe were never addressed, at least to my knowledge.

† Of course, the campaigns tended to go to the press crying foul whenever their opponent picked up votes, and they tended to be silent when errors were resolved in their favor: classic cherry-picking of data.

these errors were corrected by the time the recount ended, but many—the number is probably in the hundreds—were not.

Counting errors can't be prevented; they can be reduced somewhat, but they can never be eliminated completely. Even under the absolute best conditions imaginable, counting and tabulation errors still occur. The Minnesota voting data prove it. Errors exist no matter how small the number you're counting. People can make errors counting on their own fingers. Okay, the error rate is small and they might not do it very often, but they do it, and reasonably often if they do the same task over and over again all day. (Witness how often people will hand you the wrong amount of change because they counted out pennies incorrectly.) Some of the precincts in which errors were found were quite small; one where the audit mismatched the recount only had about 260 voters.

Minnesota's electoral law requires that there be a post-election audit. Before the election, a few hundred precincts around the state are randomly selected* to perform a hand recount of their votes to make sure that all the counting equipment was working properly. The law dictates that these recounts must use the same procedure as a full-on recount in the case of a close election. Multiple observers go through the votes as carefully and deliberately as humanly possible, trying as hard as they can to ensure that they make no errors at all. This is the ne plus ultra of counting ballots by hand. Then, because the Minnesota Senate race was so close, the very same precincts had to participate in the recount, counting those ballots again—by

---

* Their high-tech randomization device: drawing little slips from brown paper bags. Minnesota secretary of state Mark Ritchie was a bit embarrassed about the procedure, but said that the state had given up the tried-and-true fishbowl method when one official got his hand stuck in the bowl.

hand, as carefully and as deliberately as humanly possible, again ensuring that they make no errors. In these precincts, the same ultra-precise counting procedure was repeated twice. And even under these stringent conditions, the numbers didn't match up perfectly.

The errors were small, to be sure; everyone did a marvelous job preventing them. But errors there were. In the roughly two hundred precincts subject to the post-election audit, about fifty votes changed hands between the two counts. This meant that about forty-thousandths of a percent—0.04 percent—of the votes in the audit had changed between one ultra-careful count and another ultra-careful count.* This level of error wouldn't be a problem if Coleman were beating Franken by 1 percent or even half of a percent. But when the margin between the two candidates is seven-thousandths of a percent—0.007 percent—suddenly even this tiny, tiny error rate looms large. Throw on top of that the errors caused by missing or double-counted votes, issues with absentee ballots, and miscellaneous tabulation errors, and it becomes clear: these errors are at least comparable in size to—and are probably even bigger than—the number of votes that separate the two candidates. And that's even before the lizard people arrived to spirit votes away.

The unavoidable errors inherent to counting ballots would have been enough to cast doubt upon the outcome of the election. What made the election really contentious was the way partisans on both sides maneuvered to try to get their opponents' votes thrown out.

* If you restrict the numbers to the case where none of the ballots were challenged by either party, the error rate drops by roughly a factor of two, to 0.02 percent, but that underlying error rate is as low as one can go. Counting errors are unavoidable.

When the recount began, roughly two weeks after the election, the myth of "Minnesota nice" was still alive. Ramsey County's elections manager, Joe Mansky, kicked off the recount in his county by explaining the procedure for the recount: election workers would tally ballots for Franken and Coleman, while two partisan observers, one Republican and one Democrat, would observe. If either observer disagreed with how a ballot was counted, he could issue a "challenge," and the ballot would be labeled and stored in an envelope. These challenges, Mansky said, should be exceedingly rare. "I've been before the state canvassing board seventeen times," he announced. "During that time period, I think there may have been ten challenged ballots go to them . . . and the reason for that is that most Minnesotans know how to mark our ballots properly, and even the very few who don't put their mark so close to the candidate's name that there is absolutely no question about who they have voted for." The surrounding crowd—elections officials, partisan observers, and the press—nodded in agreement. Everybody was being very neighborly.

It lasted less than an hour. Within minutes, challenges were piling up. At one table, sparks were flying. A mustachioed observer in a plaid flannel shirt and voluminous pants cinched tight with a drawstring challenged ballots on absurd pretexts, hoping to disenfranchise voters who supported the Democrat. At one point, he challenged several ballots where the little Scantron ovals were clearly bubbled in for Al Franken. But, he argued, the mark was a tiny shade lighter than the other marks on the page, rendering them invalid.

Despite election officials' best efforts to keep a lid on the number of challenged ballots, the numbers kept rising and rising all around the state. Both campaigns were responsible; for every pro-Coleman knucklehead (like the one in Ramsey County) making frivolous challenges, there was a pro-Franken dimwit doing exactly the same thing (like one in Stearns County who was such a nuisance that the county issued a press release about her antics). "Both sides are behaving badly," said Mark Halvorson, an election observer from the nonpartisan Citizens for Election Integrity Minnesota. "One starts and the other escalates." By the end of the recount—even after both sides had withdrawn some of the more ludicrous claims—each side had challenged more than three thousand ballots, each of which had to go before the canvassing board for final adjudication.

The canvassing board, composed of four judges along with the Minnesota secretary of state, was flabbergasted. With nearly seven thousand challenged votes, it would have taken several weeks of mind-numbing effort to go through them all. And the vast majority of these challenges were completely bogus—a total waste of time. In hopes of speeding up the process, the canvassing board begged the campaigns to withdraw the sillier challenges. But both sides were reluctant to withdraw their challenges. There was a cynical reason behind this: proofiness.

Every evening at 9 p.m. Minnesota time, the state released the results of each day's recount—but the challenged ballots were removed from the official tallies. This meant that the campaigns could manipulate the totals with frivolous challenges: every time a Franken operative challenged a Coleman ballot, it would reduce Coleman's tally by one vote, and vice versa. Both campaigns recognized what an opportunity this presented. If the Franken campaign challenged enough ballots, they could reduce Coleman's official tally enough to make it look as if Franken were "officially" in the lead. Coleman's campaign, on the other hand, would simply challenge enough ballots to prevent that from happening, ensuring that Coleman never appeared to lose his lead.* Because of this manipulation, the official tallies of votes were all but meaningless, giving license for the campaigns to make up whatever Potemkin numbers they liked. When the recounting drew to a close, Franken's campaign declared victory—by a margin of four votes—at the same time that Coleman's staff announced a Republican victory.

The press, always fond of calling a horse race, were happy to report whatever Potemkin numbers fell into their hands. Unsophisticated reporters simply reported the official recount tallies without even trying to account for the challenged ballots. ("The Republican incumbent held a slight edge, with a 192-vote lead over Democrat Franken," reported the Associated Press.) Everyone else whipped out their crystal balls or attempted to read turkey entrails or used whatever other methods they could to generate their own projections. Poll watcher and sports statistician Nate Silver used regression analyses to predict that Franken would win by twenty-seven votes. Two weeks later, he reversed himself, projecting that Coleman would win the recount.

Of course, all of these numbers were completely worthless. The frivolous challenges ensured that the official tallies were devoid of meaning. There was no way anyone could know whether Franken or Coleman was ahead until the canvassing board finally started wading through the challenged ballots and discarding all the frivolous challenges. Yet nobody was willing to wait. Pundits and jour-

---

* This motivation became blindingly obvious as the two campaigns grudgingly withdrew their sillier challenges. Coleman's campaign wouldn't budge until Franken's campaign withdrew some challenges first. Then, when Franken's campaign withdrew a few challenges, the Coleman campaign would withdraw almost exactly the same number. This was the only way the Coleman campaign could ensure that they wouldn't lose their pseudo-lead in the vote tallies.

nalists created Potemkin numbers to titillate their audiences—those phony figures allowed them to give an answer to the unanswerable question of who was winning the race. More disturbingly, the campaigns used Potemkin numbers to foster a belief that their candidate was winning—and that their opponent was trying to steal the election. It was naked proofiness, manufactured to sway public opinion.

Not every challenge had a cynical motive. While the vast majority were frivolous, plenty of them were perfectly reasonable. Scores of ballots were challenged for valid reasons: double votes, illegible votes, bubbling-in errors, identifying marks that revealed who cast a particular ballot (one voter even got a ballot notarized).* Every possible ballot pathology—every single way a voter can conceivably mess up a ballot—was on display.

At times the fallout was stupendously absurd, as in the lizard people fiasco. The race was so close that the judges fought a hard battle for a full five minutes before coming to their decision. Mr. Lizard People might indeed be a real person, therefore the ballot—which had been counted for the Democrat—had to be discarded. Lizard People had eaten a vote for Franken.†

---

* The "no identifying marks" rule was particularly troublesome. Quite a number of people filled out the wrong circle on their ballot in pen. When they discovered the mistake, they put an X through the wrong candidate's oval and carefully filled in the oval of their true vote. These hapless voters then put their initials on the side of the ballot to designate that they approved the change—as they would on a contract—thereby putting their vote in jeopardy because they had inscribed an identifying mark.

† For the record, I think that the lizard people decision was correct; it was an overvote. However, I believe that overstrict interpretation of the statute forced the pro-Coleman forces into a ridiculous lie—to get their way, they had to pretend that they honestly believed "Lizard People" was a living, breathing individual who had somehow convinced a Minnesotan to put him up for the Senate.

The lizard people ballot was just one of the hundreds of perversities that Minnesota voters created for the amusement and confusion of the canvassing board. There were flying spaghetti monsters and Mickey Mice, ballots that had been initialed, stamped, and scrawled on in every conceivable way. There were dark circles that were nowhere near any blank ovals intended to receive the votes. A handful of voting marks had been placed with almost laserlike precision directly midway between the Coleman and Franken ovals. Mansky was right: most Minnesotans *do* know how to fill out their ballots properly. However, the ones who didn't sure failed spectacularly.

In a population of a few million voters, it's almost guaranteed that a few hundred will cast pathological ballots—ones where the voter's intent is obscure or even impossible to determine. (Badly designed ballots can make this number even worse.) This is yet another source of error, no matter how foolproof your ballots are and how strict and clear your election laws and regulations might be. Even the most careful workers will make errors when trying to interpret the meaning of a ballot, because some voters won't express their intent clearly. As a result, some ballots will be interpreted wrongly. So, in any vote that uses ballots, there are going to be ballot interpretation errors as well as counting errors. These two sources of error alone would be enough to mask the true victor of an election that is as incredibly close as the Minnesota race was. But this race had more surprises in store. There was yet another source of error that was going to make the election even uglier—and put the stop-at-nothing hypocrisy of the candidates on full display.

Every time there's a dispute about a very close election, the candidates behave in the same way. Logic forbids them from doing otherwise. The candidate who's behind by a handful of votes has to figure out some way to make up the difference between his tally and his opponent's and get into the lead. There aren't very many ways of doing this. You can try to get officials to toss out a block of votes that favor your opponent. Conversely, you can plump for including a rejected block of votes that favor you. Politicians seldom take the first tack; trying to disenfranchise voters makes you look slimier than usual. However, the second strategy is very attractive. The moment a politician is behind in a contested election, he makes high-minded speeches about how every vote in a great democracy must be counted. As he waves the flag and make orotund pronouncements about justice and fairness and the blood of our founding fathers, almost nobody notices that he's acting in his own very selfish interest when he tries to get those votes counted. And this is precisely what Al Franken's lawyers did. "There is no more precious right in a democracy than casting your vote and having it counted," Franken lawyer David Lillehaug bloviated on the day before the start of the recount. He urged the canvassing board to tally a large number of absentee ballots that had been rejected by officials. Count every vote!

For the candidate who's in the lead, the strategy has to be exactly the opposite; he has to undermine the challenger's champion-of-the-disenfranchised posturing. He declares victory based upon the first returns and tries to make it look as if the opponent were trying to steal the election by bending the rules of a fair election. He fights any of the challenger's attempts to tally more votes, because more votes (unless he knows they're favorable) have the potential to eradicate his fragile lead. Hence, Coleman's team immediately had to attempt to block the extra absentee ballots from being counted—and they had to do it without making Coleman seem like a complete ass. Not easy. "It's actually the responsibility of the voter to make sure that the vote goes where it's supposed to go," scoffed Coleman counsel Fritz Knaak when asked about the disenfranchised absentee voters. A more sympathetic lawyer would have spun the issue a little better: he would have portrayed Coleman as the law-and-order candidate, who simply holds the rule of law as sacred. It's no shame to be strict about rules and regulations, and if your candidate is trying to count dubious ballots, strictness is your friend. If a person didn't get his improperly cast vote counted, it's a shame, but rules are rules, after all.

Those were the ideological positions at the beginning of the recount, and as a result, the absentee ballots became a political and legal football—and Minnesota was swimming in absentee ballots. In the 2008 election, there were about 300,000 absentee votes—roughly 10 percent of the votes received. About 12,000 were rejected for a variety of reasons. Sometimes those reasons were valid. Sometimes the ballot came from a voter who wasn't registered. Sometimes the voter didn't sign the envelope containing the ballot or, even more commonly, failed to get a witness to sign the envelope; both signatures are required by Minnesota law. But quite a few of those 12,000 rejected ballots, numbering in the hundreds to thousands, were rejected improperly. For example, some counties rejected ballots where the voter didn't date the envelope as well as sign it, and there's nothing in electoral law that makes that vote invalid. There were a number of clear, well-documented cases of absentee ballots that had been similarly rejected in error.

As the trailing candidate, Franken and his team fought tirelessly, suing and pleading and filing papers to try to get those wrongly rejected ballots counted as soon as possible. After all, it's crucial to count every vote. The five-person canvassing board was sympathetic to their argument; these were valid votes, after all. However, a quirk in legal language meant that the canvassing board was powerless to correct the error. (The board was only able to correct errors having to do with votes *cast* during the elections, and rejected absentee ballots hadn't technically been cast.) Stretching its powers to the limit, the board requested that the counties voluntarily go through their rejected absentee ballots and pick out the ones that hadn't been rejected for a valid reason.

As the leading candidate, Coleman and his team fought tirelessly, suing and pleading and filing papers to try to uphold the rule of law by keeping those ballots out of the recount. The team tried to get a restraining order to force the issue. Coleman's obvious self-interest in blocking these absentee ballots from getting counted was beside the point. He was motivated by profound concern about the sanctity of our electoral process. Rules are rules, after all. The battle went through the court system, which issued a ruling in early January that led to some 933 extra absentee ballots being counted. But in the meantime, the landscape had shifted underneath everybody's feet.

The preliminary skirmishes over absentee ballots took place in November and December, while the recount was still under way. At the end of the recount, the five-person canvassing board plowed through all the challenged ballots, upholding the valid ones and dismissing frivolous ones. The number jumped around, but as the canvassing board's work drew to a close at the end of December, it stabilized as a nearly fifty-vote lead for Franken. For the first time, Coleman was the underdog, and Franken was the leader. And then when those 933 absentee ballots were counted, they wound up being overwhelmingly in favor of Franken. When the smoke cleared in early January, Franken was up by 225 votes.

All of a sudden, the candidates' roles had reversed. Franken was in the lead; he now merely had to sit on the football and wait. Coleman, on the other hand—the smell of desperation lingered over his camp. There weren't many paths left that could lead to a Coleman victory. The standard "count every vote" strategy wouldn't work; since uncounted ballots, particularly absentee ballots, seemed to favor Franken, counting more of them would make matters worse. So Coleman contested the election and tried to add some 650 extra ballots to the count, ballots that he admitted, in his own court filings, "were cast in areas which favored Coleman." In other words, he wanted to count only the ballots that were likely to be for him. It was electoral cherry-picking. However, even Coleman's lawyers recognized that this wasn't a winning strategy, so a few weeks later Coleman moved to plan B: trying to get thousands and thousands of other absentee ballots, including ones that were validly rejected, included in the count.

The trial went on for weeks, and it was the end of March before the outcome was clear. The court allowed an additional 400 ballots to be counted. Unfortunately for Coleman, these ballots, like the others, favored Franken and raised his lead to 312. Coleman's contest of the election was denied; Franken was declared the winner.

This wasn't the end of the saga. Coleman appealed the decision, and the Republican Party collectively crossed its fingers. If Franken managed to capture the seat, the Democrats would have sixty votes in the Senate, precisely the number they needed to break any filibuster and force through legislation over the heads of their Republican opponents. The stakes were very high, and Coleman's case looked as if it might go all the way up to the highest court in the land, the Supreme Court. If it did, *Coleman v. Franken* could well become a decision as controversial as one eight years prior: *Bush v. Gore.*

The 2000 Florida election was a nastier, dirtier, more error-ridden, and higher-stakes version of Minnesota's Senate race. As Clinton left office, the nation was deeply divided; as a result, the race to determine his successor was exceptionally close. The margin between George W. Bush and Al Gore was a bit less than half a percent of the votes cast. In some states, it was even closer. In Wisconsin, Gore's margin was less than a quarter of a percent; the same was true of Oregon. In New Mexico, Gore won by less than 0.1 percent. But those were blowouts compared to Florida.

Nobody predicted just how close the election would be in Florida. The two candidates would be separated by a mere 1,784 votes out of nearly six million, a margin of 0.03 percent, thirty-thousandths of a percent. Within a few days, an automatic recount triggered by Florida law brought the number down to 327: 0.006 percent, six-thousandths of a percent of the votes cast. On this tiny, tiny handful of votes hung the fate of the 2000 presidential election, as the two candidates had divvied up the other states of the Union more or less evenly. Florida's twenty-five electors would be decisive; whoever won Florida would take the White House.*

The 2000 presidential election was a complex drama with a rich

---

* In the United States, presidential elections have a curious structure. Citizens vote to get "electors" to represent them in a body—the electoral college—that formally determines who will be the next president. The rules of how to appoint those electors are left to the states. Most states have a winner-take-all rule; whoever gets the most votes in the election (usually) gets all of the state electors' votes. As a result, in a close race, the votes of a relatively large number of electors—even enough to determine the winner of a contest—can hinge on a few hundred votes. This is what happened in Florida.

Florida at current ocean levels.

Weiss and Overpeck, University of Arizona

Florida if the oceans rise by one meter (3.3 feet); black regions are underwater.

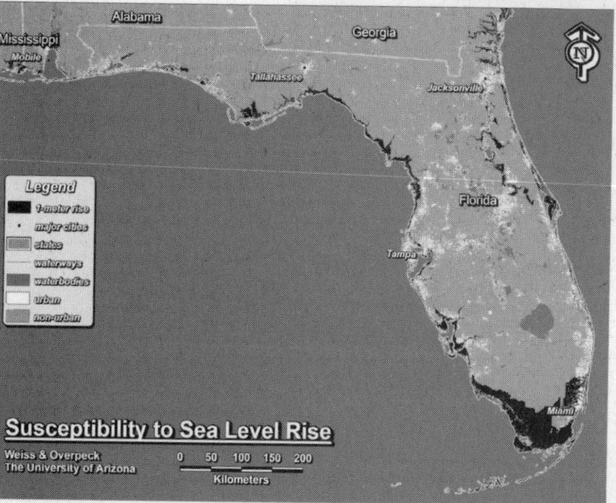

Weiss and Overpeck, University of Arizona

Florida if the oceans rise by six meters (19.7 feet).

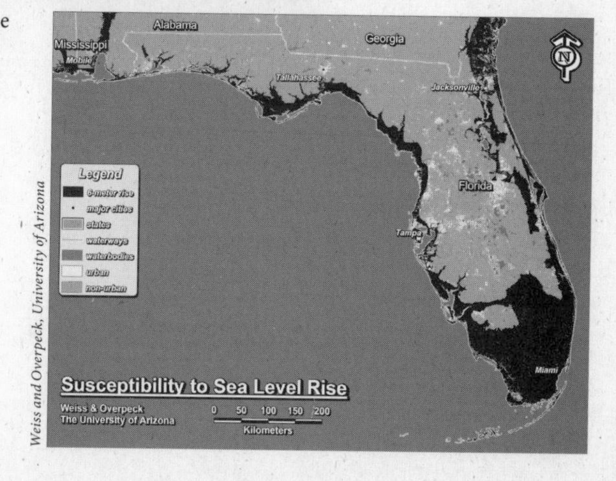

Weiss and Overpeck, University of Arizona

AP Photo/Steve Helber

On October 16, 1995, about 400,000 people gathered in Washington, D.C.—and declared themselves to be the Million Man March.

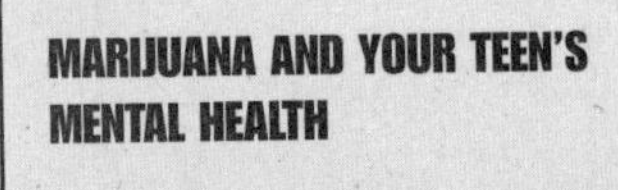

Office of National Drug Control Policy

An ad implying that marijuana use causes suicide and schizophrenia in children.

AP Photo/Byron Rollins

A triumphant Truman mocks the *Chicago Daily Tribune*.

OFFICIAL BALLOT Judge Judge

BELTRAMI COUNTY | STATE OF MINNESOTA | NOVEMBER 4, 2008

STATE GENERAL ELECTION BALLOT

INSTRUCTIONS TO VOTERS
To vote, completely fill in the oval(s) next to your choice(s) like this:

FEDERAL OFFICES

PRESIDENT AND VICE-PRESIDENT — VOTE FOR ONE TEAM
- JOHN MCCAIN AND SARAH PALIN — Republican
- BARACK OBAMA AND JOE BIDEN — Democratic-Farmer-Labor
- CYNTHIA MCKINNEY AND ROSA CLEMENTE — Green
- ROGER CALERO AND ALYSON KENNEDY — Socialist Workers
- RALPH NADER AND MATT GONZALEZ — Independent
- BOB BARR AND WAYNE A. ROOT — Libertarian
- CHUCK BALDWIN AND DARRELL CASTLE — Constitution
- Lizard People — write-in, if any

U.S. SENATOR — VOTE FOR ONE
- DEAN BARKLEY — Independence
- NORM COLEMAN — Republican
- AL FRANKEN — Democratic-Farmer-Labor
- CHARLES ALDRICH — Libertarian
- JAMES NIEMACKL — Constitution
- Lizard People — write-in, if any

U.S. REPRESENTATIVE DISTRICT 7 — VOTE FOR ONE
- GLEN MENZE — Republican
- COLLIN C. PETERSON — Democratic-Farmer-Labor
- Lizard people — write-in, if any

STATE OFFICES

STATE REPRESENTATIVE DISTRICT 4A — VOTE FOR ONE
- SHARATIN BLAKE — Independence
- JOHN CARLSON — Republican
- JOHN PERSELL — Democratic-Farmer-Labor
- Lizard People — write-in, if any

CONSTITUTIONAL AMENDMENT

Failure to vote on a constitutional amendment will have the same effect as voting no for the amendment.

To vote for a proposed constitutional amendment, completely fill in the oval next to the word "YES" for that question. To vote against a proposed constitutional amendment, completely fill in the oval next to the word "NO" for that question.

CLEAN WATER, WILDLIFE, CULTURAL HERITAGE AND NATURAL AREAS

Shall the Minnesota Constitution be amended to dedicate funding to protect our drinking water sources; to protect, enhance, and restore our wetlands, prairies, forests, and fish, game, and wildlife habitat; to preserve our arts and cultural heritage; to support our parks and trails; and to protect, enhance, and restore our lakes, rivers, streams, and groundwater by increasing the sales and use tax rate beginning July 1, 2009, by three-eighths of one percent on taxable sales until the year 2034?
- YES
- NO

COUNTY OFFICES

SOIL AND WATER CONSERVATION DISTRICT SUPERVISOR DISTRICT 3 — VOTE FOR ONE
- RUTH ANN TRASK
- Lizard people — write-in, if any

SOIL AND WATER CONSERVATION DISTRICT SUPERVISOR DISTRICT 4 — VOTE FOR ONE
- JAY BACKSTROM
- Lizard people — write-in, if any

CITY OFFICES

MAYOR CITY OF BEMIDJI — VOTE FOR ONE
- NANCY L. ERICKSON
- RICHARD LEHMANN
- Lizard People — write-in, if any

SCHOOL DISTRICT OFFICES

SCHOOL BOARD MEMBER INDEPENDENT SCHOOL DISTRICT NO. 31 (BEMIDJI) — VOTE FOR UP TO THREE
- JOHN B. PUGLEASA
- BONNIE ROCK
- ANN LONG VOELKNER
- GENE DILLON
- DAVE L. WANNER
- JAY T. HUSEBY
- LISA M. BOULAY
- write-in, if any Lizard people
- write-in, if any
- write-in, if any

0025 CITY OF BEMIDJI WARD 2 SD 31

VOTE FRONT AND BACK OF BALLOT

STATE GENERAL ELECTION BALLOT

BELTRAMI COUNTY | STATE OF MINNESOTA | NOVEMBER 4, 2008

INSTRUCTIONS TO VOTERS
To vote, completely fill in the oval(s) next to your choice(s) like this:

SCHOOL DISTRICT QUESTIONS SPECIAL ELECTION

To vote for a question, completely fill in the oval next to the word "YES" for that question. To vote against a question, completely fill in the oval next to the word "NO" for that question.

SCHOOL DISTRICT BALLOT QUESTION 1
RENEWAL OF SCHOOL DISTRICT REFERENDUM REVENUE AUTHORIZATION

The existing levy referendum authority for Independent School District No. 31, Bemidji, is expiring after 2008. The School Board has proposed to renew its expiring general education revenue of $501 per pupil. The proposed referendum revenue authorization would be applicable for five years unless otherwise revoked or reduced as provided by law.

Shall the renewal of the referendum revenue authorization proposed by the Board of Independent School District No. 31 be approved?
- YES
- NO

BY VOTING "YES" ON THIS BALLOT QUESTION, YOU ARE VOTING TO EXTEND AN EXISTING PROPERTY TAX REFERENDUM THAT IS SCHEDULED TO EXPIRE

U.S. Senator
Precinct: Ward 2
Challenged ballot #: 1
Challenged by COLEMAN
Challenge reason: Write in candidate has not been erased. Voter Intent Not Clear

JUDICIAL OFFICES

SUPREME COURT

ASSOCIATE JUSTICE 3 — VOTE FOR ONE
- TIM TINGELSTAD
- PAUL H. ANDERSON — Incumbent
- write-in, if any

ASSOCIATE JUSTICE 4 — VOTE FOR ONE
- DEBORAH HEDLUND
- LORIE SKJERVEN GILDEA — Incumbent
- write-in, if any

COURT OF APPEALS

JUDGE 16 — VOTE FOR ONE
- TERRI J. STONEBURNER — Incumbent
- DAN GRIFFITH
- write-in, if any

JUDGE 1 — VOTE FOR ONE
- EDWARD TOUSSAINT, JR. — Incumbent
- write-in, if any

JUDGE 8 — VOTE FOR ONE
- THOMAS J. KALITOWSKI — Incumbent
- write-in, if any

JUDGE 9 — VOTE FOR ONE
- ROGER M. KLAPHAKE — Incumbent
- write-in, if any

JUDGE 10 — VOTE FOR ONE
- HARRIET LANSING — Incumbent
- write-in, if any

JUDGE 15 — VOTE FOR ONE
- KEVIN G. ROSS — Incumbent
- write-in, if any

9TH DISTRICT COURT

JUDGE 3 — VOTE FOR ONE
- JOHN R. SOLIEN — Incumbent
- write-in, if any

9TH DISTRICT COURT

JUDGE 4 — VOTE FOR ONE
- DAVID F. HARRINGTON — Incumbent
- write-in, if any

JUDGE 6 — VOTE FOR ONE
- CHARLES H. (CHAD) LEDUC — Incumbent
- write-in, if any

JUDGE 8 — VOTE FOR ONE
- DONALD J. AANDAL — Incumbent
- write-in, if any

JUDGE 11 — VOTE FOR ONE
- MICHAEL J. KRAKER — Incumbent
- write-in, if any

JUDGE 12 — VOTE FOR ONE
- LOIS J. LANG — Incumbent
- write-in, if any

JUDGE 14 — VOTE FOR ONE
- ROBERT D. TIFFANY — Incumbent
- write-in, if any

JUDGE 15 — VOTE FOR ONE
- TAMARA L. YON — Incumbent
- write-in, if any

JUDGE 18 — VOTE FOR ONE
- JEFFREY S. REMICK — Incumbent
- write-in, if any

JUDGE 21 — VOTE FOR ONE
- DAVID J. TEN EYCK — Incumbent
- write-in, if any

JUDGE 22 — VOTE FOR ONE
- KURT J. MARBEN — Incumbent
- write-in, if any

VOTE FRONT AND BACK OF BALLOT

Office of the Minnesota Secretary of State

The infamous "Lizard People" ballot from Beltrami County, front (*left*) and back (*right*).

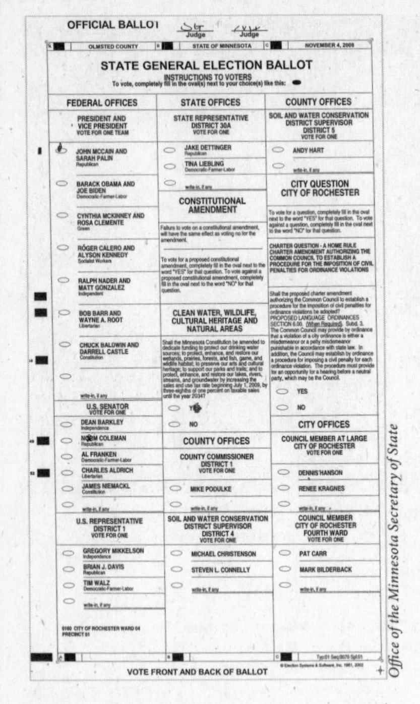

OFFICIAL BALLOT
Judge Judge

OLMSTED COUNTY | STATE OF MINNESOTA | NOVEMBER 4, 2008

STATE GENERAL ELECTION BALLOT
INSTRUCTIONS TO VOTERS
To vote, completely fill in the oval(s) next to your choice(s) like this:

FEDERAL OFFICES

PRESIDENT AND VICE PRESIDENT
VOTE FOR ONE TEAM
JOHN MCCAIN AND SARAH PALIN — Republican
BARACK OBAMA AND JOE BIDEN — Democratic-Farmer-Labor
CYNTHIA MCKINNEY AND ROSA CLEMENTE — Green
RÓGER CALERO AND ALYSON KENNEDY — Socialist Workers
RALPH NADER AND MATT GONZALEZ — Independent
BOB BARR AND WAYNE A. ROOT — Libertarian
CHUCK BALDWIN AND DARRELL CASTLE — Constitution
write-in, if any

U.S. SENATOR
VOTE FOR ONE
DEAN BARKLEY — Independence
NORM COLEMAN — Republican
AL FRANKEN — Democratic-Farmer-Labor
CHARLES ALDRICH — Libertarian
JAMES NIEMACKL — Constitution
write-in, if any

U.S. REPRESENTATIVE
DISTRICT 1
VOTE FOR ONE
GREGORY MIKKELSON — Independence
BRIAN J. DAVIS — Republican
TIM WALZ — Democratic-Farmer-Labor
write-in, if any

0160 CITY OF ROCHESTER WARD 04
PRECINCT 01

STATE OFFICES

STATE REPRESENTATIVE
DISTRICT 30A
VOTE FOR ONE
JAKE DETTINGER — Republican
TINA LIEBLING — Democratic-Farmer-Labor
write-in, if any

CONSTITUTIONAL AMENDMENT

Failure to vote on a constitutional amendment, will have the same effect as voting no for the amendment.

To vote for a proposed constitutional amendment, completely fill in the oval next to the word "YES" for that question. To vote against a proposed constitutional amendment, completely fill in the oval next to the word "NO" for that question.

CLEAN WATER, WILDLIFE, CULTURAL HERITAGE AND NATURAL AREAS

Shall the Minnesota Constitution be amended to dedicate funding to protect our drinking water sources; to protect, enhance, and restore our wetlands, prairies, forests, and fish, game, and wildlife habitat; to preserve our arts and cultural heritage; to support our parks and trails; and to protect, enhance, and restore our lakes, rivers, streams, and groundwater by increasing the sales and use tax rate beginning July 1, 2009, by three-eighths of one percent on taxable sales until the year 2034?
YES
NO

COUNTY OFFICES

COUNTY COMMISSIONER
DISTRICT 1
VOTE FOR ONE
MIKE PODULKE
write-in, if any

SOIL AND WATER CONSERVATION DISTRICT SUPERVISOR
DISTRICT 4
VOTE FOR ONE
MICHAEL CHRISTENSON
STEVEN L. CONNELLY
write-in, if any

COUNTY OFFICES

SOIL AND WATER CONSERVATION DISTRICT SUPERVISOR
DISTRICT 5
VOTE FOR ONE
ANDY HART
write-in, if any

CITY QUESTION
CITY OF ROCHESTER

To vote for a question, completely fill in the oval next to the word "YES" for that question. To vote against a question, completely fill in the oval next to the word "NO" for that question.

CHARTER QUESTION - A HOME RULE CHARTER AMENDMENT AUTHORIZING THE COMMON COUNCIL TO ESTABLISH A PROCEDURE FOR THE IMPOSITION OF CIVIL PENALTIES FOR ORDINANCE VIOLATIONS

Shall the proposed charter amendment authorizing the Common Council to establish a procedure for the imposition of civil penalties for ordinance violations be adopted?
PROPOSED LANGUAGE ORDINANCES
SECTION 6.00. (When Required). Subd. 3. The Common Council may provide by ordinance that a violation of a city ordinance is either a misdemeanor or a petty misdemeanor punishable in accordance with state law. In addition, the Council may establish by ordinance a procedure for imposing a civil penalty for each ordinance violation. The procedure must provide for an opportunity for a hearing before a neutral party, which may be the Council.
YES
NO

CITY OFFICES

COUNCIL MEMBER AT LARGE
CITY OF ROCHESTER
VOTE FOR ONE
DENNIS HANSON
RENEE KRAGNES
write-in, if any

COUNCIL MEMBER
CITY OF ROCHESTER
FOURTH WARD
VOTE FOR ONE
PAT CARR
MARK BILDERBACK
write-in, if any

Typ:01 Seq:0070 Spl:01
VOTE FRONT AND BACK OF BALLOT
© Election Systems & Software, Inc. 1981, 2002

Office of the Minnesota Secretary of State

OFFICIAL BALLOT
Judge
Judge

STATE GENERAL ELECTION BALLOT
DAKOTA COUNTY, MINNESOTA
NOVEMBER 4, 2008

INSTRUCTIONS TO VOTERS
To vote, completely fill in the oval(s) next to your choice(s) like this:

FEDERAL OFFICES

PRESIDENT AND VICE-PRESIDENT
VOTE FOR ONE TEAM
JOHN MCCAIN AND SARAH PALIN — Republican
BARACK OBAMA AND JOE BIDEN — Democratic-Farmer-Labor
CYNTHIA MCKINNEY AND ROSA CLEMENTE — Green
ROGER CALERO AND ALYSON KENNEDY — Socialist Workers
RALPH NADER AND MATT GONZALEZ — Independent
BOB BARR AND WAYNE A. ROOT — Libertarian
CHUCK BALDWIN AND DARRELL CASTLE — Constitution
write-in, if any

UNITED STATES SENATOR
VOTE FOR ONE
DEAN BARKLEY — Independence
NORM COLEMAN — Republican
AL FRANKEN — Democratic-Farmer-Labor
CHARLES ALDRICH — Libertarian
JAMES NIEMACKL — Constitution
write-in, if any

UNITED STATES REPRESENTATIVE
DISTRICT 2
VOTE FOR ONE
JOHN KLINE — Republican
STEVE SARVI — Democratic-Farmer-Labor
write-in, if any

STATE OFFICES

STATE REPRESENTATIVE
DISTRICT 40A
VOTE FOR ONE
TODD JOHNSON — Republican
WILL MORGAN — Democratic-Farmer-Labor
write-in, if any

CONSTITUTIONAL AMENDMENT

Failure to vote on a constitutional amendment, will have the same effect as voting no for the amendment.

To vote for a proposed constitutional amendment, completely fill in the oval next to the word "YES" for that question. To vote against a proposed constitutional amendment, completely fill in the oval next to the word "NO" for that question.

CLEAN WATER, WILDLIFE, CULTURAL HERITAGE, AND NATURAL AREAS

Shall the Minnesota Constitution be amended to dedicate funding to protect our drinking water sources; to protect, enhance, and restore our wetlands, prairies, forests, and fish, game, and wildlife habitat; to preserve our arts and cultural heritage; to support our parks and trails; and to protect, enhance, and restore our lakes, rivers, streams, and groundwater by increasing the sales and use tax rate beginning July 1, 2009, by three-eighths of one percent on taxable sales until the year 2034?
YES
NO

COUNTY OFFICES

COUNTY COMMISSIONER
DISTRICT 5
VOTE FOR ONE
LIZ WORKMAN
VICKY TURNER
write-in, if any

SOIL AND WATER CONSERVATION DISTRICT SUPERVISOR
DISTRICT 2
VOTE FOR ONE
MARIAN BROWN
VICTORIA A. DVORAK
SCOTT NORSTAD
write-in, if any

SOIL AND WATER CONSERVATION DISTRICT SUPERVISOR
DISTRICT 4
VOTE FOR ONE
CHRIS NIELSEN
PETER THOMAS SCHAFFER
write-in, if any

SOIL AND WATER CONSERVATION DISTRICT SUPERVISOR
DISTRICT 5
VOTE FOR ONE
JOE MEYERS
MICHAEL T. CARR

CITY OFFICES
CITY OF BURNSVILLE

MAYOR
VOTE FOR ONE
ELIZABETH B. KAUTZ
JERRY WILLENBURG
write-in, if any

COUNCIL MEMBER
VOTE FOR UP TO TWO
MICHAEL ESCH
DAN GUSTAFSON
MARY SHERRY
write-in, if any
write-in, if any

Office of the Minnesota Secretary of State

Above, left: Marking the "o" in "Norm."

Above, right: Sitting on the fence between Franken and Coleman.

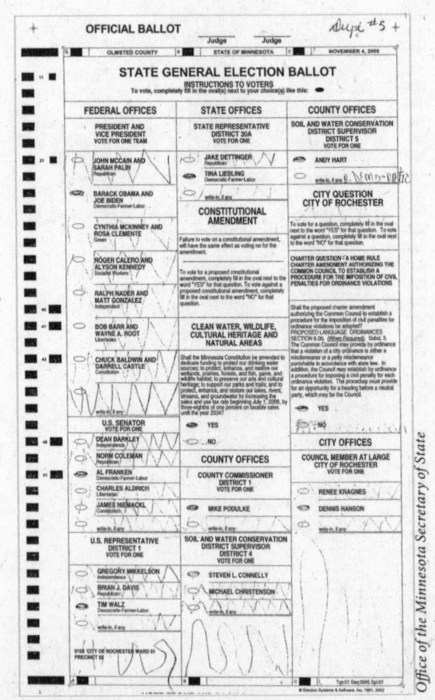

OFFICIAL BALLOT
Dupe #5
Judge Judge

OLMSTED COUNTY | STATE OF MINNESOTA | NOVEMBER 4, 2008

STATE GENERAL ELECTION BALLOT
INSTRUCTIONS TO VOTERS
To vote, completely fill in the oval(s) next to your choice(s) like this:

FEDERAL OFFICES

PRESIDENT AND VICE PRESIDENT
VOTE FOR ONE TEAM
JOHN MCCAIN AND SARAH PALIN — Republican
BARACK OBAMA AND JOE BIDEN — Democratic-Farmer-Labor
CYNTHIA MCKINNEY AND ROSA CLEMENTE — Green
RÓGER CALERO AND ALYSON KENNEDY — Socialist Workers
RALPH NADER AND MATT GONZALEZ — Independent
BOB BARR AND WAYNE A. ROOT — Libertarian
CHUCK BALDWIN AND DARRELL CASTLE — Constitution
write-in, if any

U.S. SENATOR
VOTE FOR ONE
DEAN BARKLEY — Independence
NORM COLEMAN — Republican
AL FRANKEN — Democratic-Farmer-Labor
CHARLES ALDRICH — Libertarian
JAMES NIEMACKL — Constitution
write-in, if any

U.S. REPRESENTATIVE
DISTRICT 1
VOTE FOR ONE
GREGORY MIKKELSON — Independence
BRIAN J. DAVIS — Republican
TIM WALZ — Democratic-Farmer-Labor
write-in, if any

0105 CITY OF ROCHESTER WARD 01
PRECINCT 02

STATE OFFICES

STATE REPRESENTATIVE
DISTRICT 30A
VOTE FOR ONE
JAKE DETTINGER — Republican
TINA LIEBLING — Democratic-Farmer-Labor
write-in, if any

CONSTITUTIONAL AMENDMENT

Failure to vote on a constitutional amendment, will have the same effect as voting no for the amendment.

To vote for a proposed constitutional amendment, completely fill in the oval next to the word "YES" for that question. To vote against a proposed constitutional amendment, completely fill in the oval next to the word "NO" for that question.

CLEAN WATER, WILDLIFE, CULTURAL HERITAGE AND NATURAL AREAS

Shall the Minnesota Constitution be amended to dedicate funding to protect our drinking water sources; to protect, enhance, and restore our wetlands, prairies, forests, and fish, game, and wildlife habitat; to preserve our arts and cultural heritage; to support our parks and trails; and to protect, enhance, and restore our lakes, rivers, streams, and groundwater by increasing the sales and use tax rate beginning July 1, 2009, by three-eighths of one percent on taxable sales until the year 2034?
YES
NO

COUNTY OFFICES

COUNTY COMMISSIONER
DISTRICT 1
VOTE FOR ONE
MIKE PODULKE
write-in, if any

SOIL AND WATER CONSERVATION DISTRICT SUPERVISOR
DISTRICT 4
VOTE FOR ONE
STEVEN L. CONNELLY
MICHAEL CHRISTENSON
write-in, if any

COUNTY OFFICES

SOIL AND WATER CONSERVATION DISTRICT SUPERVISOR
DISTRICT 5
VOTE FOR ONE
ANDY HART
write-in, if any

CITY QUESTION
CITY OF ROCHESTER

To vote for a question, completely fill in the oval next to the word "YES" for that question. To vote against a question, completely fill in the oval next to the word "NO" for that question.

CHARTER QUESTION - A HOME RULE CHARTER AMENDMENT AUTHORIZING THE COMMON COUNCIL TO ESTABLISH A PROCEDURE FOR THE IMPOSITION OF CIVIL PENALTIES FOR ORDINANCE VIOLATIONS

Shall the proposed charter amendment authorizing the Common Council to establish a procedure for the imposition of civil penalties for ordinance violations be adopted?
PROPOSED LANGUAGE ORDINANCES
SECTION 6.00. (When Required). Subd. 3. The Common Council may provide by ordinance that a violation of a city ordinance is either a misdemeanor or a petty misdemeanor punishable in accordance with state law. In addition, the Council may establish by ordinance a procedure for imposing a civil penalty for each ordinance violation. The procedure must provide for an opportunity for a hearing before a neutral party, which may be the Council.
YES
NO

CITY OFFICES

COUNCIL MEMBER AT LARGE
CITY OF ROCHESTER
VOTE FOR ONE
RENEE KRAGNES
DENNIS HANSON
write-in, if any

Typ:01 Seq:0045 Spl:01
© Election Systems & Software, Inc. 1981, 2002

Office of the Minnesota Secretary of State

Voting with too much emphasis.

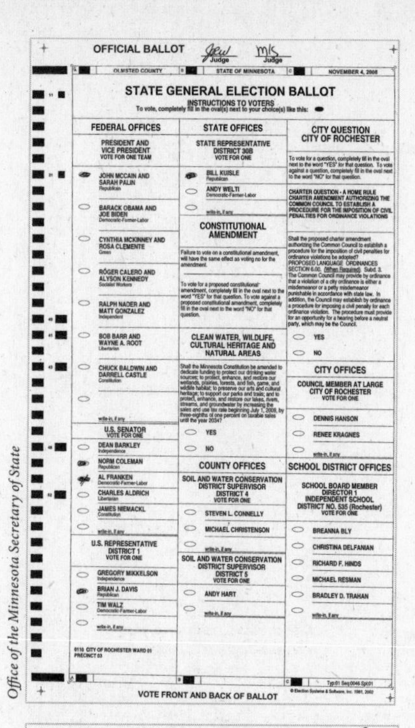

OFFICIAL BALLOT — Judge — Judge

OLMSTED COUNTY | STATE OF MINNESOTA | NOVEMBER 4, 2008

STATE GENERAL ELECTION BALLOT

INSTRUCTIONS TO VOTERS
To vote, completely fill in the oval(s) next to your choice(s) like this:

FEDERAL OFFICES

PRESIDENT AND VICE PRESIDENT
VOTE FOR ONE TEAM

- JOHN MCCAIN AND SARAH PALIN — Republican
- BARACK OBAMA AND JOE BIDEN — Democratic-Farmer-Labor
- CYNTHIA MCKINNEY AND ROSA CLEMENTE — Green
- RÓGER CALERO AND ALYSON KENNEDY — Socialist Workers
- RALPH NADER AND MATT GONZALEZ — Independent
- BOB BARR AND WAYNE A. ROOT — Libertarian
- CHUCK BALDWIN AND DARRELL CASTLE — Constitution
- write-in, if any

U.S. SENATOR
VOTE FOR ONE

- DEAN BARKLEY — Independence
- NORM COLEMAN — Republican
- AL FRANKEN — Democratic-Farmer-Labor
- CHARLES ALDRICH — Libertarian
- JAMES NIEMACKL — Constitution
- write-in, if any

U.S. REPRESENTATIVE
DISTRICT 1
VOTE FOR ONE

- GREGORY MIKKELSON — Independence
- BRIAN J. DAVIS — Republican
- TIM WALZ — Democratic-Farmer-Labor
- write-in, if any

0110 CITY OF ROCHESTER WARD 01
PRECINCT 03

STATE OFFICES

STATE REPRESENTATIVE
DISTRICT 30B
VOTE FOR ONE

- BILL KUISLE — Republican
- ANDY WELTI — Democratic-Farmer-Labor
- write-in, if any

CONSTITUTIONAL AMENDMENT

Failure to vote on a constitutional amendment, will have the same effect as voting no for the amendment.

To vote for a proposed constitutional amendment, completely fill in the oval next to the word "YES" for that question. To vote against a proposed constitutional amendment, completely fill in the oval next to the word "NO" for that question.

CLEAN WATER, WILDLIFE, CULTURAL HERITAGE AND NATURAL AREAS

Shall the Minnesota Constitution be amended to dedicate funding to protect our drinking water sources; to protect, enhance, and restore our wetlands, prairies, forests, and fish, game, and wildlife habitat; to preserve our arts and cultural heritage; to support our parks and trails; and to protect, enhance, and restore our lakes, rivers, streams, and groundwater by increasing the sales and use tax rate beginning July 1, 2009, by three-eighths of one percent on taxable sales until the year 2034?

- YES
- NO

COUNTY OFFICES

SOIL AND WATER CONSERVATION DISTRICT SUPERVISOR
DISTRICT 4
VOTE FOR ONE

- STEVEN L. CONNELLY
- MICHAEL CHRISTENSON
- write-in, if any

SOIL AND WATER CONSERVATION DISTRICT SUPERVISOR
DISTRICT 5
VOTE FOR ONE

- ANDY HART
- write-in, if any

CITY QUESTION
CITY OF ROCHESTER

To vote for a question, completely fill in the oval next to the word "YES" for that question. To vote against a question, completely fill in the oval next to the word "NO" for that question.

CHARTER QUESTION - A HOME RULE CHARTER AMENDMENT AUTHORIZING THE COMMON COUNCIL TO ESTABLISH A PROCEDURE FOR THE IMPOSITION OF CIVIL PENALTIES FOR ORDINANCE VIOLATIONS

Shall the proposed charter amendment authorizing the Common Council to establish a procedure for the imposition of civil penalties for ordinance violations be adopted?
PROPOSED LANGUAGE ORDINANCES SECTION 6.00. (When Required). Subd. 3. The Common Council may provide by ordinance that a violation of a city ordinance is either a misdemeanor or a petty misdemeanor punishable in accordance with state law. In addition, the Council may establish by ordinance a procedure for imposing a civil penalty for each ordinance violation. The procedure must provide for an opportunity for a hearing before a neutral party, which may be the Council.

- YES
- NO

CITY OFFICES

COUNCIL MEMBER AT LARGE
CITY OF ROCHESTER
VOTE FOR ONE

- DENNIS HANSON
- RENEE KRAGNES
- write-in, if any

SCHOOL DISTRICT OFFICES

SCHOOL BOARD MEMBER
DIRECTOR 1
INDEPENDENT SCHOOL DISTRICT NO. 535 (Rochester)
VOTE FOR ONE

- BREANNA BLY
- CHRISTINA DELFANIAN
- RICHARD F. HINDS
- MICHAEL RESMAN
- BRADLEY D. TRAHAN
- write-in, if any

Typ:01 Seq:0046 Spl:01
© Election Systems & Software, Inc. 1981, 2002

VOTE FRONT AND BACK OF BALLOT

Office of the Minnesota Secretary of State

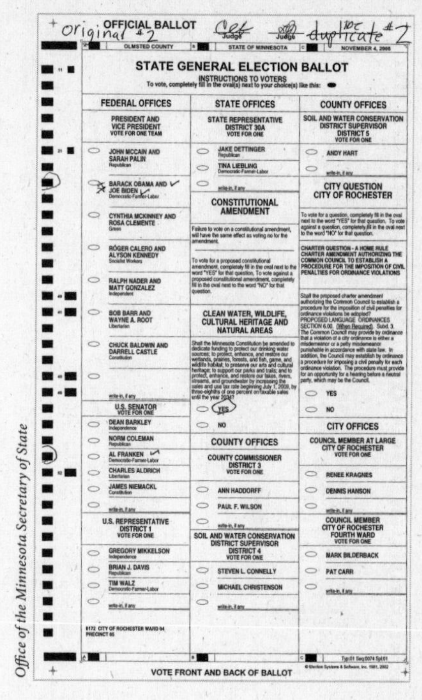

original #2 — OFFICIAL BALLOT — Judge — Judge — duplicate #2

OLMSTED COUNTY | STATE OF MINNESOTA | NOVEMBER 4, 2008

STATE GENERAL ELECTION BALLOT

INSTRUCTIONS TO VOTERS
To vote, completely fill in the oval(s) next to your choice(s) like this:

FEDERAL OFFICES

PRESIDENT AND VICE PRESIDENT
VOTE FOR ONE TEAM

- JOHN MCCAIN AND SARAH PALIN — Republican
- BARACK OBAMA AND JOE BIDEN — Democratic-Farmer-Labor
- CYNTHIA MCKINNEY AND ROSA CLEMENTE — Green
- RÓGER CALERO AND ALYSON KENNEDY — Socialist Workers
- RALPH NADER AND MATT GONZALEZ — Independent
- BOB BARR AND WAYNE A. ROOT — Libertarian
- CHUCK BALDWIN AND DARRELL CASTLE — Constitution
- write-in, if any

U.S. SENATOR
VOTE FOR ONE

- DEAN BARKLEY — Independence
- NORM COLEMAN — Republican
- AL FRANKEN — Democratic-Farmer-Labor
- CHARLES ALDRICH — Libertarian
- JAMES NIEMACKL — Constitution
- write-in, if any

U.S. REPRESENTATIVE
DISTRICT 1
VOTE FOR ONE

- GREGORY MIKKELSON — Independence
- BRIAN J. DAVIS — Republican
- TIM WALZ — Democratic-Farmer-Labor
- write-in, if any

0172 CITY OF ROCHESTER WARD 04
PRECINCT 05

STATE OFFICES

STATE REPRESENTATIVE
DISTRICT 30A
VOTE FOR ONE

- JAKE DETTINGER — Republican
- TINA LIEBLING — Democratic-Farmer-Labor
- write-in, if any

CONSTITUTIONAL AMENDMENT

Failure to vote on a constitutional amendment, will have the same effect as voting no for the amendment.

To vote for a proposed constitutional amendment, completely fill in the oval next to the word "YES" for that question. To vote against a proposed constitutional amendment, completely fill in the oval next to the word "NO" for that question.

CLEAN WATER, WILDLIFE, CULTURAL HERITAGE AND NATURAL AREAS

Shall the Minnesota Constitution be amended to dedicate funding to protect our drinking water sources; to protect, enhance, and restore our wetlands, prairies, forests, and fish, game, and wildlife habitat; to preserve our arts and cultural heritage; to support our parks and trails; and to protect, enhance, and restore our lakes, rivers, streams, and groundwater by increasing the sales and use tax rate beginning July 1, 2009, by three-eighths of one percent on taxable sales until the year 2034?

- YES
- NO

COUNTY OFFICES

COUNTY COMMISSIONER
DISTRICT 3
VOTE FOR ONE

- ANN HADDORFF
- PAUL F. WILSON
- write-in, if any

SOIL AND WATER CONSERVATION DISTRICT SUPERVISOR
DISTRICT 4
VOTE FOR ONE

- STEVEN L. CONNELLY
- MICHAEL CHRISTENSON
- write-in, if any

SOIL AND WATER CONSERVATION DISTRICT SUPERVISOR
DISTRICT 5
VOTE FOR ONE

- ANDY HART
- write-in, if any

CITY QUESTION
CITY OF ROCHESTER

To vote for a question, completely fill in the oval next to the word "YES" for that question. To vote against a question, completely fill in the oval next to the word "NO" for that question.

CHARTER QUESTION - A HOME RULE CHARTER AMENDMENT AUTHORIZING THE COMMON COUNCIL TO ESTABLISH A PROCEDURE FOR THE IMPOSITION OF CIVIL PENALTIES FOR ORDINANCE VIOLATIONS

Shall the proposed charter amendment authorizing the Common Council to establish a procedure for the imposition of civil penalties for ordinance violations be adopted?
PROPOSED LANGUAGE ORDINANCES SECTION 6.00. (When Required). Subd. 3. The Common Council may provide by ordinance that a violation of a city ordinance is either a misdemeanor or a petty misdemeanor punishable in accordance with state law. In addition, the Council may establish by ordinance a procedure for imposing a civil penalty for each ordinance violation. The procedure must provide for an opportunity for a hearing before a neutral party, which may be the Council.

- YES
- NO

CITY OFFICES

COUNCIL MEMBER AT LARGE
CITY OF ROCHESTER
VOTE FOR ONE

- RENEE KRAGNES
- DENNIS HANSON
- write-in, if any

COUNCIL MEMBER
CITY OF ROCHESTER
FOURTH WARD
VOTE FOR ONE

- MARK BILDERBACK
- PAT CARR
- write-in, if any

Typ:01 Seq:0074 Spl:01
© Election Systems & Software, Inc. 1981, 2002

VOTE FRONT AND BACK OF BALLOT

Office of the Minnesota Secretary of State

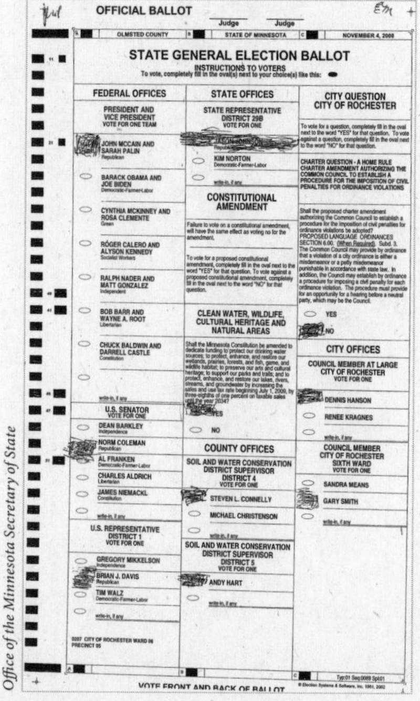

OFFICIAL BALLOT — Judge — Judge

OLMSTED COUNTY | STATE OF MINNESOTA | NOVEMBER 4, 2008

STATE GENERAL ELECTION BALLOT

INSTRUCTIONS TO VOTERS
To vote, completely fill in the oval(s) next to your choice(s) like this:

FEDERAL OFFICES

PRESIDENT AND VICE PRESIDENT
VOTE FOR ONE TEAM

- JOHN MCCAIN AND SARAH PALIN — Republican
- BARACK OBAMA AND JOE BIDEN — Democratic-Farmer-Labor
- CYNTHIA MCKINNEY AND ROSA CLEMENTE — Green
- RÓGER CALERO AND ALYSON KENNEDY — Socialist Workers
- RALPH NADER AND MATT GONZALEZ — Independent
- BOB BARR AND WAYNE A. ROOT — Libertarian
- CHUCK BALDWIN AND DARRELL CASTLE — Constitution
- write-in, if any

U.S. SENATOR
VOTE FOR ONE

- DEAN BARKLEY — Independence
- NORM COLEMAN — Republican
- AL FRANKEN — Democratic-Farmer-Labor
- CHARLES ALDRICH — Libertarian
- JAMES NIEMACKL — Constitution
- write-in, if any

U.S. REPRESENTATIVE
DISTRICT 1
VOTE FOR ONE

- GREGORY MIKKELSON — Independence
- BRIAN J. DAVIS — Republican
- TIM WALZ — Democratic-Farmer-Labor
- write-in, if any

0207 CITY OF ROCHESTER WARD 06
PRECINCT 05

STATE OFFICES

STATE REPRESENTATIVE
DISTRICT 29B
VOTE FOR ONE

- JASON JOHNSON — Republican
- KIM NORTON — Democratic-Farmer-Labor
- write-in, if any

CONSTITUTIONAL AMENDMENT

Failure to vote on a constitutional amendment, will have the same effect as voting no for the amendment.

To vote for a proposed constitutional amendment, completely fill in the oval next to the word "YES" for that question. To vote against a proposed constitutional amendment, completely fill in the oval next to the word "NO" for that question.

CLEAN WATER, WILDLIFE, CULTURAL HERITAGE AND NATURAL AREAS

Shall the Minnesota Constitution be amended to dedicate funding to protect our drinking water sources; to protect, enhance, and restore our wetlands, prairies, forests, and fish, game, and wildlife habitat; to preserve our arts and cultural heritage; to support our parks and trails; and to protect, enhance, and restore our lakes, rivers, streams, and groundwater by increasing the sales and use tax rate beginning July 1, 2009, by three-eighths of one percent on taxable sales until the year 2034?

- YES
- NO

COUNTY OFFICES

SOIL AND WATER CONSERVATION DISTRICT SUPERVISOR
DISTRICT 4
VOTE FOR ONE

- STEVEN L. CONNELLY
- MICHAEL CHRISTENSON
- write-in, if any

SOIL AND WATER CONSERVATION DISTRICT SUPERVISOR
DISTRICT 5
VOTE FOR ONE

- ANDY HART
- write-in, if any

CITY QUESTION
CITY OF ROCHESTER

To vote for a question, completely fill in the oval next to the word "YES" for that question. To vote against a question, completely fill in the oval next to the word "NO" for that question.

CHARTER QUESTION - A HOME RULE CHARTER AMENDMENT AUTHORIZING THE COMMON COUNCIL TO ESTABLISH A PROCEDURE FOR THE IMPOSITION OF CIVIL PENALTIES FOR ORDINANCE VIOLATIONS

Shall the proposed charter amendment authorizing the Common Council to establish a procedure for the imposition of civil penalties for ordinance violations be adopted?
PROPOSED LANGUAGE ORDINANCES SECTION 6.00. (When Required). Subd. 3. The Common Council may provide by ordinance that a violation of a city ordinance is either a misdemeanor or a petty misdemeanor punishable in accordance with state law. In addition, the Council may establish by ordinance a procedure for imposing a civil penalty for each ordinance violation. The procedure must provide for an opportunity for a hearing before a neutral party, which may be the Council.

- YES
- NO

CITY OFFICES

COUNCIL MEMBER AT LARGE
CITY OF ROCHESTER
VOTE FOR ONE

- DENNIS HANSON
- RENEE KRAGNES
- write-in, if any

COUNCIL MEMBER
CITY OF ROCHESTER
SIXTH WARD
VOTE FOR ONE

- SANDRA MEANS
- GARY SMITH
- write-in, if any

Typ:01 Seq:0089 Spl:01
© Election Systems & Software, Inc. 1981, 2002

VOTE FRONT AND BACK OF BALLOT

Office of the Minnesota Secretary of State

Above, left: A crossed-out vote . . . maybe.

Above, right: A circle? A checkmark? Both are wrong.

A vote for Coleman invades Franken territory.

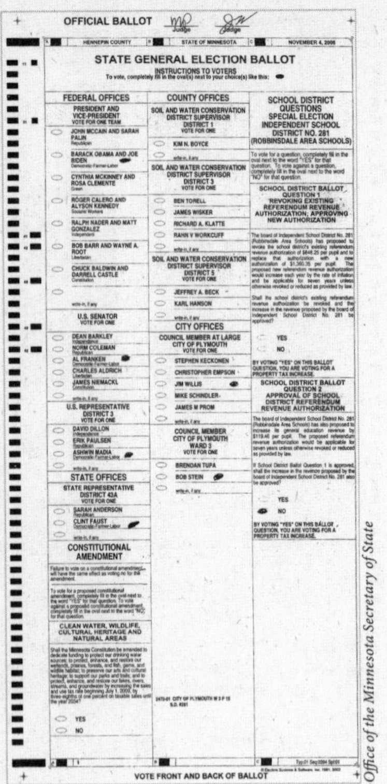

OFFICIAL BALLOT — Judge — Judge

HENNEPIN COUNTY | STATE OF MINNESOTA | NOVEMBER 4, 2008

STATE GENERAL ELECTION BALLOT

INSTRUCTIONS TO VOTERS
To vote, completely fill in the oval(s) next to your choice(s) like this:

FEDERAL OFFICES

PRESIDENT AND VICE-PRESIDENT
VOTE FOR ONE TEAM

- JOHN MCCAIN AND SARAH PALIN — Republican
- BARACK OBAMA AND JOE BIDEN — Democratic-Farmer-Labor
- CYNTHIA MCKINNEY AND ROSA CLEMENTE — Green
- ROGER CALERO AND ALYSON KENNEDY — Socialist Workers
- RALPH NADER AND MATT GONZALEZ — Independent
- BOB BARR AND WAYNE A. ROOT — Libertarian
- CHUCK BALDWIN AND DARRELL CASTLE — Constitution
- write-in, if any

U.S. SENATOR
VOTE FOR ONE

- DEAN BARKLEY — Independence
- NORM COLEMAN — Republican
- AL FRANKEN — Democratic-Farmer-Labor
- CHARLES ALDRICH — Libertarian
- JAMES NIEMACKL — Constitution
- write-in, if any

U.S. REPRESENTATIVE DISTRICT 3
VOTE FOR ONE

- DAVID DILLON — Independence
- ERIK PAULSEN — Republican
- ASHWIN MADIA — Democratic-Farmer-Labor
- write-in, if any

STATE OFFICES

STATE REPRESENTATIVE DISTRICT 43A
VOTE FOR ONE

- SARAH ANDERSON — Republican
- CLINT FAUST — Democratic-Farmer-Labor
- write-in, if any

CONSTITUTIONAL AMENDMENT

Failure to vote on a constitutional amendment will have the same effect as voting no for the amendment.

To vote for a proposed constitutional amendment, completely fill in the oval next to the word "YES" for that question. To vote against a proposed constitutional amendment, completely fill in the oval next to the word "NO" for that question.

CLEAN WATER, WILDLIFE, CULTURAL HERITAGE AND NATURAL AREAS

Shall the Minnesota Constitution be amended to dedicate funding to protect our drinking water sources; to protect, enhance, and restore our wetlands, prairies, forests, and fish, game, and wildlife habitat; to preserve our arts and cultural heritage; to support our parks and trails; and to protect, enhance, and restore our lakes, rivers, streams, and groundwater by increasing the sales and use tax rate beginning July 1, 2009, by three-eighths of one percent on taxable sales until the year 2034?

- YES
- NO

COUNTY OFFICES

SOIL AND WATER CONSERVATION DISTRICT SUPERVISOR DISTRICT 1
VOTE FOR ONE

- KIM N. BOYCE
- write-in, if any

SOIL AND WATER CONSERVATION DISTRICT SUPERVISOR DISTRICT 3
VOTE FOR ONE

- BEN TORELL
- JAMES WISKER
- RICHARD A. KLATTE
- RAHN V WORKCUFF
- write-in, if any

SOIL AND WATER CONSERVATION DISTRICT SUPERVISOR DISTRICT 5
VOTE FOR ONE

- JEFFREY A. BECK
- KARL HANSON
- write-in, if any

CITY OFFICES

COUNCIL MEMBER AT LARGE CITY OF PLYMOUTH
VOTE FOR ONE

- STEPHEN KECKONEN
- CHRISTOPHER EMPSON
- JIM WILLIS
- MIKE SCHINDLER
- JAMES M PROM
- write-in, if any

COUNCIL MEMBER CITY OF PLYMOUTH WARD 3
VOTE FOR ONE

- BRENDAN TUPA
- BOB STEIN
- write-in, if any

2470-01 CITY OF PLYMOUTH W 3 P 15 S.D. #281

SCHOOL DISTRICT QUESTIONS
SPECIAL ELECTION
INDEPENDENT SCHOOL DISTRICT NO. 281
(ROBBINSDALE AREA SCHOOLS)

To vote for a question, completely fill in the oval next to the word "YES" for that question. To vote against a question, completely fill in the oval next to the word "NO" for that question.

SCHOOL DISTRICT BALLOT QUESTION 1
REVOKING EXISTING REFERENDUM REVENUE AUTHORIZATION; APPROVING NEW AUTHORIZATION

The board of Independent School District No. 281 (Robbinsdale Area Schools) has proposed to revoke the school district's existing referendum revenue authorization of $848.25 per pupil and to replace that authorization with a new authorization of $1,360.35 per pupil. The proposed new referendum revenue authorization would increase each year by the rate of inflation and be applicable for seven years unless otherwise revoked or reduced as provided by law.

Shall the school district's existing referendum revenue authorization be revoked and the increase in the revenue proposed by the board of Independent School District No. 281 be approved?

- YES
- NO

BY VOTING "YES" ON THIS BALLOT QUESTION, YOU ARE VOTING FOR A PROPERTY TAX INCREASE.

SCHOOL DISTRICT BALLOT QUESTION 2
APPROVAL OF SCHOOL DISTRICT REFERENDUM REVENUE AUTHORIZATION

The board of Independent School District No. 281 (Robbinsdale Area Schools) has also proposed to increase its general education revenue by $119.46 per pupil. The proposed referendum revenue authorization would be applicable for seven years unless otherwise revoked or reduced as provided by law.

If School District Ballot Question 1 is approved, shall the increase in the revenue proposed by the board of Independent School District No. 281 also be approved?

- YES
- NO

BY VOTING "YES" ON THIS BALLOT QUESTION, YOU ARE VOTING FOR A PROPERTY TAX INCREASE.

Typ:01 Seq:0394 Spl:01
© Election Systems & Software, Inc. 1981, 2002

VOTE FRONT AND BACK OF BALLOT

Office of the Minnesota Secretary of State

Franken gets at least one far-right vote.

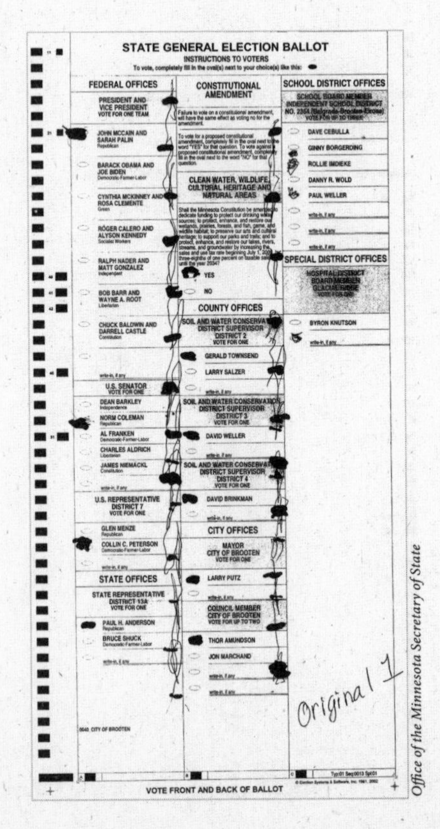

STATE GENERAL ELECTION BALLOT

INSTRUCTIONS TO VOTERS
To vote, completely fill in the oval(s) next to your choice(s) like this:

FEDERAL OFFICES

PRESIDENT AND VICE PRESIDENT
VOTE FOR ONE TEAM

- JOHN MCCAIN AND SARAH PALIN — Republican
- BARACK OBAMA AND JOE BIDEN — Democratic-Farmer-Labor
- CYNTHIA MCKINNEY AND ROSA CLEMENTE — Green
- ROGER CALERO AND ALYSON KENNEDY — Socialist Workers
- RALPH NADER AND MATT GONZALEZ — Independent
- BOB BARR AND WAYNE A. ROOT — Libertarian
- CHUCK BALDWIN AND DARRELL CASTLE — Constitution
- write-in, if any

U.S. SENATOR
VOTE FOR ONE

- DEAN BARKLEY — Independence
- NORM COLEMAN — Republican
- AL FRANKEN — Democratic-Farmer-Labor
- CHARLES ALDRICH — Libertarian
- JAMES NIEMACKL — Constitution
- write-in, if any

U.S. REPRESENTATIVE DISTRICT 7
VOTE FOR ONE

- GLEN MENZE — Republican
- COLLIN C. PETERSON — Democratic-Farmer-Labor
- write-in, if any

STATE OFFICES

STATE REPRESENTATIVE DISTRICT 13A
VOTE FOR ONE

- PAUL H. ANDERSON — Republican
- BRUCE SHUCK — Democratic-Farmer-Labor
- write-in, if any

0640 CITY OF BROOTEN

CONSTITUTIONAL AMENDMENT

Failure to vote on a constitutional amendment, will have the same effect as voting no for the amendment.

To vote for a proposed constitutional amendment, completely fill in the oval next to the word "YES" for that question. To vote against a proposed constitutional amendment, completely fill in the oval next to the word "NO" for that question.

CLEAN WATER, WILDLIFE, CULTURAL HERITAGE AND NATURAL AREAS

Shall the Minnesota Constitution be amended to dedicate funding to protect our drinking water sources; to protect, enhance, and restore our wetlands, prairies, forests, and fish, game, and wildlife habitat; to preserve our arts and cultural heritage; to support our parks and trails; and to protect, enhance, and restore our lakes, rivers, streams, and groundwater by increasing the sales and use tax rate beginning July 1, 2009, by three-eighths of one percent on taxable sales until the year 2034?

- YES
- NO

COUNTY OFFICES

SOIL AND WATER CONSERVATION DISTRICT SUPERVISOR DISTRICT 2
VOTE FOR ONE

- GERALD TOWNSEND
- LARRY SALZER
- write-in, if any

SOIL AND WATER CONSERVATION DISTRICT SUPERVISOR DISTRICT 3
VOTE FOR ONE

- DAVID WELLER
- write-in, if any

SOIL AND WATER CONSERVATION DISTRICT SUPERVISOR DISTRICT 4
VOTE FOR ONE

- DAVID BRINKMAN
- write-in, if any

CITY OFFICES

MAYOR CITY OF BROOTEN
VOTE FOR ONE

- LARRY PUTZ
- write-in, if any

COUNCIL MEMBER CITY OF BROOTEN
VOTE FOR UP TO TWO

- THOR AMUNDSON
- JON MARCHAND
- write-in, if any
- write-in, if any

SCHOOL DISTRICT OFFICES

SCHOOL BOARD MEMBER INDEPENDENT SCHOOL DISTRICT NO. 2364 (Belgrade-Brooten-Elrosa)
VOTE FOR UP TO THREE

- DAVE CEBULLA
- GINNY BORGERDING
- ROLLIE IMDIEKE
- DANNY R. WOLD
- PAUL WELLER
- write-in, if any
- write-in, if any
- write-in, if any

SPECIAL DISTRICT OFFICES

HOSPITAL DISTRICT BOARD MEMBER GLACIAL RIDGE
VOTE FOR ONE

- BYRON KNUTSON
- write-in, if any

Original 1

Typ:01 Seq:0013 Spl:01
© Election Systems & Software, Inc. 1981, 2002

VOTE FRONT AND BACK OF BALLOT

Office of the Minnesota Secretary of State

The "jellyfish" ballot.

Charles Seife

Recounting ballots in Ramsey County.

Charles Seife

Pausing at the Ramsey County recount to read about the falling market.

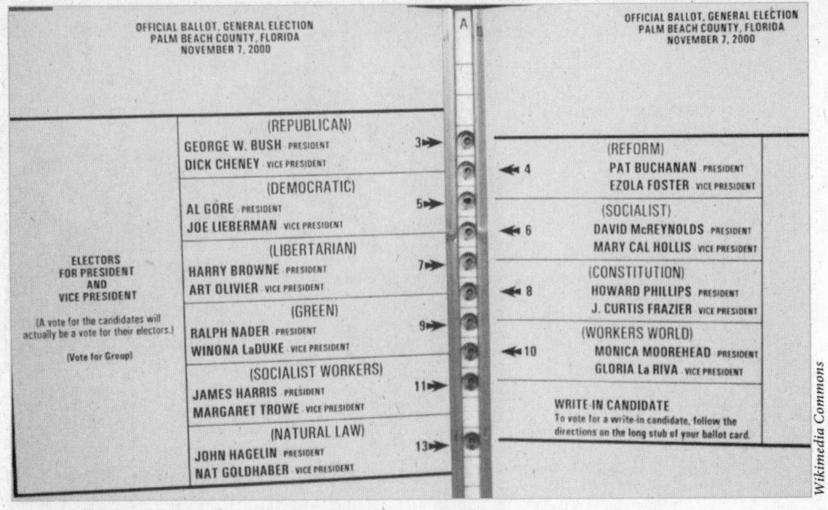

OFFICIAL BALLOT, GENERAL ELECTION
PALM BEACH COUNTY, FLORIDA
NOVEMBER 7, 2000

ELECTORS
FOR PRESIDENT
AND
VICE PRESIDENT

(A vote for the candidates will actually be a vote for their electors.)

(Vote for Group)

(REPUBLICAN)
GEORGE W. BUSH - PRESIDENT
DICK CHENEY - VICE PRESIDENT
3→

(DEMOCRATIC)
AL GORE - PRESIDENT
JOE LIEBERMAN - VICE PRESIDENT
5→

(LIBERTARIAN)
HARRY BROWNE - PRESIDENT
ART OLIVIER - VICE PRESIDENT
7→

(GREEN)
RALPH NADER - PRESIDENT
WINONA LaDUKE - VICE PRESIDENT
9→

(SOCIALIST WORKERS)
JAMES HARRIS - PRESIDENT
MARGARET TROWE - VICE PRESIDENT
11→

(NATURAL LAW)
JOHN HAGELIN - PRESIDENT
NAT GOLDHABER - VICE PRESIDENT
13→

A

OFFICIAL BALLOT, GENERAL ELECTION
PALM BEACH COUNTY, FLORIDA
NOVEMBER 7, 2000

←4
(REFORM)
PAT BUCHANAN - PRESIDENT
EZOLA FOSTER - VICE PRESIDENT

←6
(SOCIALIST)
DAVID McREYNOLDS - PRESIDENT
MARY CAL HOLLIS - VICE PRESIDENT

←8
(CONSTITUTION)
HOWARD PHILLIPS - PRESIDENT
J. CURTIS FRAZIER - VICE PRESIDENT

←10
(WORKERS WORLD)
MONICA MOOREHEAD - PRESIDENT
GLORIA La RIVA - VICE PRESIDENT

WRITE-IN CANDIDATE
To vote for a write-in candidate, follow the directions on the long stub of your ballot card.

Wikimedia Commons

The Palm Beach County butterfly ballot.

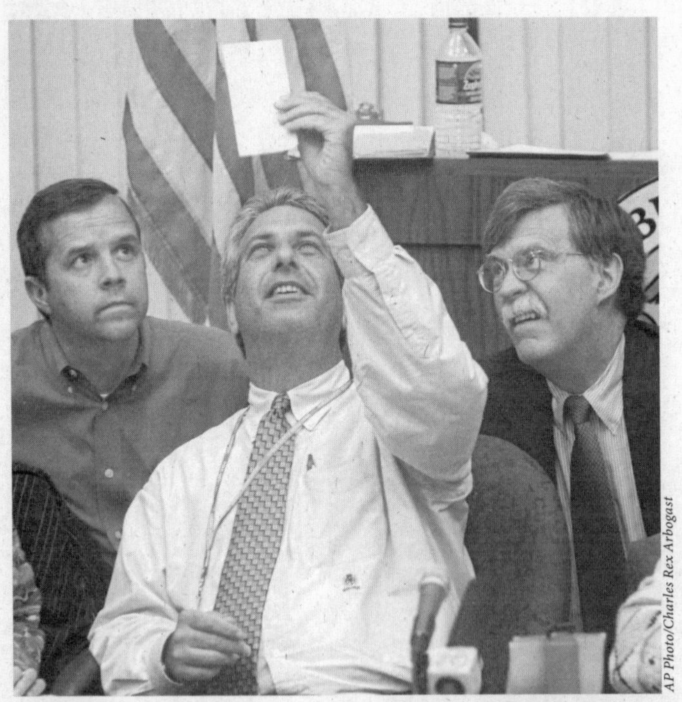

AP Photo/Charles Rex Arbogast

Judge Charles Burton demonstrating his ultraprecise method of reading disputed punchcards during the 2000 presidential election, as Democrat Mark White (*left*) and Republican John Bolton (*right*) observe.

supporting cast and ironies aplenty. But the true complexity of the drama came from the nature of Florida elections. Unlike Minnesota, the electoral laws in Florida were neither clean nor modern; they were messy, vague, and full of holes. While Minnesota had a uniform system of voting—every single precinct used relatively reliable Scantron machines to tally ballots of similar designs in a standardized manner—Florida's voting system was a patchwork. Each county had different machines, different systems, and different electoral rules. Some used punch cards while others used Scantrons. There was no standardized ballot, and the punch-card-reading machines were notoriously unreliable. As mentioned earlier, a provision in state law triggered a recount in close elections; as in Minnesota, an election with a margin less than 0.5 percent of the votes cast automatically forced counties to count the ballots again. However, the recount wasn't to be done by hand except under certain circumstances—unless there was a proven "error in the vote tabulation," ballots would simply be passed through the readers a second time.

Adding to the drama, the patchwork nature of the Florida election made it incredibly error-prone—much more so than Minnesota's. Minnesota's ballots were relatively uniform and well-designed, making it difficult for citizens to cast their vote for the wrong person (though some certainly managed). In Florida there was no uniform design, and some of the ballots were incredibly confusing. In Palm Beach County, a wealthy, Democratic-leaning region of Florida, an infamously badly designed punch-card ballot, the so-called butterfly ballot, invalidated thousands of votes for Gore. Confused by the ballot's layout, the (mostly elderly) citizens accidentally punched out the hole belonging to Pat Buchanan, a far-right candidate, instead of (or in addition to) their vote for Al Gore. It was crystal clear that this was a systematic error that cost Gore thousands of votes.*

Another source of errors was punch cards. Even in 2000, punch-card ballots were antiquated and unreliable and had large error rates. One major problem was that it was surprisingly difficult for voters to punch the cards properly. Most of the ballots were made out of a small piece of stiff paper with little rectangles scored into the sheet. The voters put the ballot into a voting machine—essentially a box with a metal grate on the top—and with a little stylus punched out rectangles ("chads") corresponding to the votes they wanted to cast. In theory, the box was supposed to ensure that the chad was completely removed from the ballot, as well as keep the voting booths tidy by storing the discarded chads. Unfortunately, the machines didn't always work as advertised. Often the chad didn't completely detach from the ballot, leaving it dangling by one or two corners and making it hard for a machine reader to interpret properly. Sometimes—especially when the box was full of other people's chads—the stylus would merely crease or dimple the chad rather than detach it, making it look like a no-vote to a machine. Punch-card systems tended to be used in poorer, generally Democratic areas, while wealthier ones tended to use Scantron ballots.

As the curtain opened on the election drama, the actors took up their assigned roles without a moment's hesitation. Gore, like Franken, was behind, so he was cast in the role of the flag-waving cham-

* The number of Buchanan votes, roughly 3,400 of them, was more than ten times the number of votes Buchanan was expected to receive given the history of the county and the voting patterns of Florida residents, even under extraordinary conditions. When Bush spokesman and later press secretary Ari Fleischer explained away the anomaly by insisting that "Palm Beach County is a Pat Buchanan stronghold," even the Buchanan staff called the claim "nonsense."

pion of the disenfranchised; he had to find votes to count, preferably from pro-Gore districts, in hopes of catching up with Bush. His eyes turned first to areas that were Gore strongholds: Volusia, Palm Beach, Miami-Dade, and Broward counties. His team argued that these four (Gore-supporting) counties, plagued by punch-card errors, must be recounted by hand rather than by machine—machines weren't able to count improperly punched ballots. At the same time, though, a hand count would likely introduce new errors. Punch cards are relatively fragile. The very act of handling them would alter a number of ballots—causing some loose chads to fall out. Unlike hardy pen-marked, bubbled-in ballots, punch-card ballots can be spoiled merely by handling them. What's more, it's not obvious how to interpret a ballot with a chad that's still partially attached. Did it represent a genuine vote, or was it an artifact of somebody's handling the ballot too roughly? These were judgment calls that almost certainly were introducing error into the count—and that partisan observers would fight over in attempts to gather votes for their candidates. The Florida courts, largely populated with Democrat-appointed judges, tended to side with Gore and his count-every-vote argument.

Bush was ahead, so he was cast in the role of spoiler. He immediately declared victory and claimed that his opponent was attempting to steal the election. (Republicans quickly branded the Gore-Lieberman ticket as "Sore-Loserman.") More important, Bush and his team had to counter Gore's maneuvers. When Gore tried to get recounts started, Bush fought in the courts to stop them dead—but he was rebuffed. Luckily for Bush, though, he had another recourse. The governor of Florida happened to be his younger brother, so the state government, which is in charge of administering the election, was full of Bush partisans. The secretary of state,

Katherine Harris, and the head of Florida's division of elections, L. Clayton Roberts, were clearly in the Bush camp; though they attempted to appear fair-minded, to all outward appearances they worked as hard as they could (within the constraints of their offices) to kill Gore's recounts dead, with great success. When Gore won in the courts, the state government mooted those victories.

The main plotline in the Bush versus Gore drama followed this pattern over and over again. All the dirty dealing, infighting, manipulation, lying, and nastiness added texture to the tale, but didn't alter its trajectory. Gore and his allies pushed to get more votes included in the tallies. (Count every vote!) Bush and his allies used deadlines and strict interpretations of laws and regulations to block those votes from getting hand-counted and, when that strategy failed, to keep the hand counts from being included in the official tally. (Rules are rules!) Gore tended to prevail in the courts, but the court victories were often made worthless by a counterattack from pro-Bush state officials. The recounts started and stopped in response to a flurry of lawsuits, judicial rulings, and election memos.

A secondary plotline showed how hypocritical both sides were. In truth, the "count every vote" ethic had not been as deeply ingrained in the Gore camp as they pretended; they abandoned the principle when it seemed to be to their advantage. Similarly, Bush abdicated his role as the strict upholder of law and order the moment it looked like he could pick up a few votes by bending the rules. When it came to absentee ballots (which tended to favor Bush), Gore's team was suddenly a stickler for rules, ensuring that absentee ballots without the required postmark were tossed. Bush's team was more sophisticated in its hypocrisy. In areas that were likely to break for Gore, they were also sticklers, tossing out ballots;

however, for military and overseas ballots, which favored Bush, the team huffed and puffed, delivering high-minded lectures on the need to count every vote, especially from our brave fighting people. "I can not believe that our service boys, fighting hard overseas, that their ballots would be disqualified," a Bush lawyer asked with all the indignation he could muster. When convenient, Bush suddenly became a "count every vote" person, just as Gore had transformed into a "rules are rules" politician.* However, these battles took place in the dark periphery of the election fight, far from center stage. Few saw how shallow both sides' convictions truly were, how hypocritical everybody was being in their attempts to grab the golden ring. The spotlight was on the battle over the punch-card ballots.

Bush won the first round of that battle, effectively blocking Gore's push for recounts. When Katherine Harris certified the official vote totals on November 26—pointedly ignoring manual recount results—Bush retained a slim lead, 2,912,790 to 2,912,253. This was a margin of 537 votes, or 0.009 percent: nine-thousandths of a percent of the votes cast. Gore immediately counterattacked. Though he lost a preliminary skirmish in court, he prevailed on appeal; the Florida Supreme Court ordered an immediate start to

---

* Another example of this role reversal came from a set of lawsuits in Seminole and Martin counties. These lawsuits hinged on allegations that Republican operatives were allowed to alter improperly filled-out ballot applications—a shady procedure that tainted thousands upon thousands of absentee ballots in these counties. The pro-Gore forces (not Gore himself, even though some of his legal team tried to convince him to join in the suit) tried to get those ballots thrown out, as Florida law would seem to dictate. (Rules are rules!) If they succeeded, the suits would likely have eliminated a net of 2,000 to 2,500 votes for Bush, handing Gore the victory. Predictably, the pro-Bush forces (including Bush's team) fought hard to keep those ballots in the contest. (Count every vote!) In this particular battle, the Bush team prevailed.

manual recounts in *every* county—not just the four pro-Gore ones—to look for votes that the tabulating machines had failed to count. Round two had gone to Gore. The third round, the battle in the Supreme Court, would determine the victor.

On December 9, the day after the Florida Supreme Court ordered the manual recounts, the U.S. Supreme Court jumped enthusiastically into the fray. It landed directly on Al Gore's neck, stopping the recounts cold. Five of the nine justices—Scalia, Thomas, Rehnquist, O'Connor, and Kennedy—argued that counting illegal votes could cause "irreparable harm" to the legitimacy of the election. The remaining four—Stevens, Ginsburg, Breyer, and Souter—argued that not completing the recount would be an even worse blow to the election's legitimacy. The conservatives won; by a five-to-four vote, they put a temporary stop to the manual recounts. A few days later, the halt became permanent. Without a hint of irony, the Court determined that the recounts (which they had halted) couldn't be completed in time to meet a deadline implied by federal law and therefore had to be abandoned entirely. Game, set, match. Bush wins.

This was the element of *Bush v. Gore* that was most controversial because it seemed nakedly partisan. However, there were other troubling parts of the decision. One section has the potential to wreak havoc on the nation's entire electoral process. It is an argument that uses one form of proofiness to come to the disturbing conclusion that performing the recount in Florida would have been a violation of the equal protection clause in the Fourteenth Amendment to the Constitution.

Briefly, the equal protection clause dictates that each citizen should be treated equally by the law. It's an important if vague principle, and it has repercussions all throughout our legal system.

In electoral law, it has been interpreted to mean that no one person's vote is valued more than another's: one person, one vote. (What *this* means is also subject to interpretation. Taken too literally, it can be used to argue that our entire system of voting based upon the electoral college is unconstitutional.)* The majority opinions in *Bush v. Gore* were generous in their interpretation of the clause and found equal protection violations everywhere. For example, as long as you have different election officials making slightly different judgment calls when interpreting the meaning of ballots, it means that some ballots—the ones in the hands of particularly lenient officials—are more likely to be counted than others.

Some of the equal protection concerns were valid. Our democracy is plagued with problems caused by votes being given different weights (more on this in the next chapter). However, the problems outlined in *Bush v. Gore* could have been remedied by coming up with a uniform, reasonable, and practical set of standards for conducting the recounts. This the Court deemed impossible. Thus the Court ruled, "It is obvious that the recount cannot be conducted with the requirements of equal protection and due process" without doing impractical and, frankly, idiotic tasks like evaluating new software for the voting machines to get them to pick out overvotes. Functionally, the Court was ruling that manual recounts of any sort where two different election judges might disagree about the intent of a ballot are unconstitutional. It was a jaw-dropping ruling that threw a huge monkey wrench into our electoral process: it essentially made all recounts—and any other electoral pro-

---

* Looking at the ratio of voters per electoral college vote as a crude measure of the worth of a vote, a vote in the District of Columbia in the 2000 presidential election was twice as powerful as a vote in California. This means, pretty obviously, that the votes aren't worth the same amount.

cesses that might not be 100 percent objective or consistent—unconstitutional.*

This is proofiness. The Supreme Court was deliberately ignoring the errors inherent in determining the victor of an electoral contest. No matter how careful we are, there are going to be mistakes in counting and tabulating ballots. Since we are human beings, we are never 100 percent consistent or objective when making judgment calls in elections. The Minnesota election proves that even in the best-designed recounts with the most prescriptive of laws, there will be counting errors, and some oddly marked ballots will cause even the most reasonable and rational people to fight about how to interpret them. There's no clearer proof that the Supreme Court's insistence on perfect objectivity is misguided. It's delusional to pretend that elections can ever be perfect. They're sloppy affairs by their very nature. We're just lucky that most of the time the outcome of an election is clear enough to be visible through the muck.

When it isn't, all bets are off. Close elections are, by their nature, vicious affairs. Candidates make noble speeches all the while trying to steal the election from their opponents with whatever lies are

---

* The majority judges recognized how damaging their equal protection argument was. They wanted to elect Bush, but not suffer the consequences of the logic that they used, so they slipped a little clause into their ruling that was intended to prevent it from ever being used as a precedent. "Our consideration is limited to the present circumstances, for the problem of equal protection in election processes generally presents many complexities." This was unprecedented; never before had the Supreme Court tried to evade the consequences of its rulings. In trying to get a freebie, one-time-only decision, the Court created legal chaos. ("Respectfully, the Supreme Court does not issue non-precedential decisions," decreed a lower court a few years later. You know things are in disarray when lower courts are dinging the Supreme Court rather than vice versa.) Merely trying to cite *Bush v. Gore* leads to a logical paradox—how can you make an argument that uses *Bush v. Gore* if the very act of citing it means that you disagree with it? (Of course, this doesn't stop lawyers from trying!)

most likely to sway the public. They use lawyers and government officials and judges to fight partisan battles. Even the most august institutions in the land—such as the Supreme Court—have given in to the temptation of bartering their credibility to assure a candidate's victory.

Even when the courts and judges behave admirably, a close election is certain to mean a tremendous waste of time and money. The state of Minnesota spent thousands of dollars during the recount, while Coleman's and Franken's campaigns spent tens of millions. Minnesota's citizens were missing a senator for eight months before the issue was resolved. It could have taken longer had Coleman appealed all the way to the Supreme Court as Gore had, but after losing his claim in Minnesota's highest court, he conceded. Eight months after the election, Minnesota no longer was shy a senator.

It doesn't have to be this way. Mathematicians have ways of dealing with uncertainty. They can take even the closest, nastiest election and come up with a way of determining the victor in a manner that's fair to all the parties involved. Instead, close elections wind up in the hands of lawyers. At that point, the outcome is essentially arbitrary. Who got more ballots is almost irrelevant to the outcome. The winner is determined by luck, by lawyers, and by proofiness.

A close election is the ideal breeding ground for proofiness of all kinds. When lawyers declared that the fluctuating vote totals in the Minnesota election were "statistically dubious"—implying that they were the product of electoral fraud—they were using randumbness to generate phony plots and conspiracies. Partisans on both sides also cherry-pick data that is favorable to the press, all the while trying to bury data that is unfavorable. For example, one oft-repeated tidbit that echoed through the press in January: "In twenty-five precincts, most of them Franken precincts, there were more votes cast than showed up at the polls." In fact, the number was closer to seven hundred precincts, and plenty of them favored Coleman—quite a significant omission.* These sorts of randumbness and cherry-picking are sins of spin; they foment discontent and try to undermine support for an opponent. They're nasty and they're dishonest, but they're not the worst of what's going on.

In a close election, one particular kind of proofiness, shared by Democrats and Republicans alike, is responsible for the worst lies in *Coleman v. Franken* as well as *Bush v. Gore*. And this form of proofiness all but guaranteed that these close elections would be decided the wrong way. As indeed they were. Only by ridding ourselves of it can we finally see who should have won the Minnesota election—or whether Bush should have ever been made president. That form of proofiness is disestimation.

Disestimation is the root of all the evils that came from the 2000 presidential election as well as the 2008 Minnesota Senate election. As long as there are close elections, disestimation will continue to exact a toll year after year, obscuring the truth. However, once you recognize the problem and see through the fog of disestimation, it becomes crystal clear how these elections *should* have been decided.

The first step is simply to admit that elections are measurements

---

* What's more, the vast majority of the big mismatches—discrepancies of more than a vote or two—were in pro-Franken Ramsey County. As mentioned earlier, a pack of raccoons was apparently in charge of keeping Ramsey County's voter turnout records; they were demonstrably and obviously kept incorrectly. The discrepancy was because the voter turnout records were unreliable, not because ballots were magically materializing out of thin air (or being double-counted).

and are therefore fallible. Even under ideal conditions, even when officials count well-designed ballots with incredible deliberation, there are errors on the order of a few hundredths of a percent. And that's just the beginning. There are plenty of other errors in any election. There are errors caused by people entering data incorrectly. There are errors created by people filling out ballots wrong, casting their vote for the wrong person. There are errors caused by absentee ballots that are wrongly excluded from the count, or illegal ballots that somehow make it into the mix. Ballots will be misplaced. Ballots will be double-counted. All of these errors make it harder to determine the truth of who really won an election.

Everybody knows about these errors, at least on some level. In *Bush v. Gore* the Court commented on how the dispute "brought into sharp focus a common, if heretofore unnoticed phenomenon"* that the process of counting ballots was error-prone. Yet almost everybody completely ignores these errors. This is disestimation.

When the Minnesota secretary of state announced that Barack Obama got 1,573,354 votes and John McCain got 1,275,409 votes in the 2008 election, it was a case of disestimation. By presenting the number in that manner, the secretary of state made it seem as if this electoral measurement was much better than it actually was. The announcement is just as silly as the proverbial museum guide's claim that a dinosaur is 65,000,038 years old. In fact, given the errors, the correct thing to say would be something like Obama won approximately 1,573,000 votes and McCain won approximately 1,275,000. This is the most accurate statement you can make about the election, given the magnitude of the errors in the tabulation process. (Even this is probably still implying too much precision.)

---

* "Heretofore unnoticed" if you've been living on Mars, that is.

However, because the margin between the candidates is so large, the point is moot.

With Franken versus Coleman, it was just as wrong for the secretary of state to declare that Franken had won 1,212,629 votes to Coleman's 1,212,317. The proper thing to say was that both Franken and Coleman had each received approximately 1,212,000 votes. The Florida 2000 election was even more error-prone and thus the official result was even more of a disestimate. It shouldn't have been Bush 2,912,790, Gore 2,912,253. In truth, the result was something like Bush, approximately 2,910,000; Gore, approximately 2,910,000. These are the most accurate statements you can make about the elections in Minnesota and Florida. It is impossible to get more precise, given the errors inherent to the measurement.

All of the angry battles over a few hundred votes here and there are pointless; they're fights over a complete fiction. The Coleman-Franken battle about whether one won by 206 votes or the other won by 312 is equivalent to two museum guides battling over whether the dinosaur is 65,000,038 or 65,000,019 years old: the outcome has no bearing on reality. All the arguments over absentee ballots, hanging chads, and the like are a complete waste of time; they're merely a side effect of the proofiness that almost everyone is complicit in. We're all complicit because nobody wants to face the truth.

And the truth is this: the Minnesota 2008 Senate race was tied. Both Franken and Coleman got the same amount of votes, at least as best as we humans can tell with the instruments we have at our disposal. The winner of the 2000 Florida presidential race? Nobody. Bush and Gore got the same number of votes, to the best of anybody's knowledge. That race too was tied. The counting methods available to us do not allow us to give any answer more precise than that, so any answer other than that is a fiction—it is mere proofiness.

Nevertheless, everybody buys into the myth that those last few hundred votes that separate Bush and Gore or Coleman and Franken are meaningful. Government officials do it because they don't want to admit that their ability to count votes is limited. Reporters do it because they crave certainty; if presented with approximate data, they'd try, against all logic, to figure out the *real* victor.* Political candidates buy into the myth too—the person in the lead has every reason to pretend that his lead is more than mere illusion, while the candidate who trails has to pretend that there's meaning in the handful of uncounted ballots that he fights for in court. All of these parties have their own reasons for clinging to the fiction of disestimation. As a result, none of the players who have the power to anoint the winner of an election—the government, the press, and especially the candidates themselves—will ever do the right thing and declare a close election to be a tie.

But that's the truth of the matter. Clear the proofiness away, and it becomes obvious that the Minnesota and Florida elections were ties. And this reveals what the outcome of those elections should have been.

The Minnesota and Florida election laws each had a procedure for what happens in a tied election. In Florida, this procedure was described in the state's statutes, title IX, chapter 103.162; in Minnesota's election laws, it was in chapter 204C.34. Cut through the verbiage and it becomes apparent that both states, by coincidence,

---

* This is what they did in *Bush v. Gore*; a number of newspapers spent oodles of money and time counting and recounting ballots in various ways. Their conclusion: Gore won . . . or Bush won, depending on how you count.

happen to break a tie in exactly the same way. In the case of a tie vote, the winner shall be determined by lot. In other words, flip a coin.

It's hard to swallow, but the 2008 Minnesota Senate race and, even more startling, the 2000 presidential election should have been settled with the flip of a coin.

6

An Unfair Vote

It's very hard in military or in personal life to assure complete equality. Life is unfair.

—John Fitzgerald Kennedy

I just received the following wire from my generous daddy: "Dear Jack, Don't buy a single vote more than is necessary. I'll be damned if I'm going to pay for a landslide."

—John Fitzgerald Kennedy

Elections are inherently unfair. No matter what method a government uses to run an election, it can't be an equal contest, at least in a mathematical sense. It's an inescapable truth: all elections are flawed, and there's nothing we can do to fix them.

That's the bad news. The *really* bad news: a number of politicians and judges are making our flawed system of elections much, much more unfair than it already is.

Democracy is inherently an institution based upon a mathematical operation—that of counting votes. As we've seen, that operation is vulnerable to proofiness. Armed with bogus mathematical arguments and underhanded tactics, politicians and their judicial allies are working to stack the electoral deck to get their party into power and keep it there. They are succeeding.

Democracy is in danger, buckling under an assault from proofiness.

At the same time that Norm Coleman's final appeal was limping around the hallways of the Minnesota Supreme Court, Minnesota justices were deciding yet another lawsuit about Minnesota elections—one that had the potential to change the way elections in Minnesota are run. In theory, it could prevent fiascos like the Franken-Coleman race from ever happening again.

As an experiment, the city of Minneapolis dispensed with the standard vote-for-one-candidate "plurality" method for deciding certain elections. Instead, the city would use what's known as instant runoff voting, where each voter would rank the candidates in order of preference. Advocates argue that instant runoff voting makes elections more fair and more transparent. They have a point: had instant runoff voting been used in the 2008 Senate election, officials would almost certainly have been able to declare a victor within a matter of days.

After the recount, after all the legal challenges, the final results of the 2008 Minnesota Senate election were:

1,212,317 for Norm Coleman
1,212,629 for Al Franken
437,505 for Dean Barkley

Since this was a plurality election, with each voter casting a single vote in the Senate race, the winner was simply the person with the most votes—Al Franken, in this case. But because the election was so close (and the lawyers were so skilled), it took eight months and millions of dollars to determine the final winner.

An instant runoff vote, on the other hand, would probably have made the recount unnecessary. In such a vote, the ballots are slightly different from what they are in a plurality election. Instead of voting for a single candidate, each voter gets to rank the candidates, from least to greatest of all possible evils. Using those ranks, officials can figure out who wins the election. It's not quite as simple as merely counting votes; an instant runoff version of the 2008 Minnesota race would have had a much more complicated-looking result than a plurality election. It might have looked something like:

1,202,310 prefer Norm Coleman over Dean Barkley over Al Franken
10,007 prefer Norm Coleman over Al Franken over Dean Barkley
1,201,620 prefer Al Franken over Dean Barkley over Norm Coleman
11,009 prefer Al Franken over Norm Coleman over Dean Barkley
287,010 prefer Dean Barkley over Norm Coleman over Al Franken
150,495 prefer Dean Barkley over Al Franken over Norm Coleman

So . . . who wins? This requires a little number juggling. First, you look at everybody's first choice. In this case:

1,212,317 chose Norm Coleman as their first choice
1,212,629 chose Al Franken as their first choice
437,505 chose Dean Barkley as their first choice

. . . exactly the same as in the plurality election. But the election isn't over yet. In an instant runoff vote, you can only win if you get more than 50 percent of the votes. Neither Franken (at 42 percent) nor Coleman (also at 42 percent) managed to cross that threshold and win a majority of votes. When this happens, the "instant runoff" begins: the candidate at the bottom of the pack, Dean Barkley, is eliminated; it becomes a two-person race between Coleman and Franken. Barkley voters aren't disenfranchised, though. Officials count their second-choice votes in lieu of the now moot first-choice votes for Barkley. Using the above—hypothetical—numbers, the result of such an instant runoff would be:

1,499,327 votes for Norm Coleman (including 287,010 former Barkley voters)
1,363,124 votes for Al Franken (including 150,495 former Barkley voters)

Norm Coleman now has a solid majority so the election ends. As an added bonus, the race is no longer terribly close—the margin is nearly 5 percent of the votes cast—so there's no need for a recount. Instant runoff voting would likely have saved the state of Minnesota a whole lot of trouble.

However, when the city of Minneapolis proposed using instant runoff voting for some elections, opponents promptly sued, claiming that the scheme is unfair—more specifically, it's unconstitutional because it doesn't count every vote in precisely the same way, an argument that the Minnesota Supreme Court promptly knocked down. However, it's absolutely true that instant runoff voting has disadvantages over plurality voting. Ballots in plurality elections are very simple—you simply vote for one person—yet as the Minnesota election showed, people screw them up all the time. Imagine all the ways that people would fill out a "rank three candidates in order of preference" ballot incorrectly. The error rate of the election would go through the roof.

Another drawback of instant runoff voting is that it doesn't always give the answer that the voting public thinks is best. Indeed, in the above example, Dean Barkley could argue that in fact he should be elected, because that choice would make the most people the least unhappy. To see this, take the instant runoff results above and modify them slightly. Every time a voter ranks a candidate in first place, give that candidate two points to represent the voter's strong desire to get that candidate elected. Every time a voter ranks a candidate second, the candidate gets one point, signifying that the candidate isn't the voter's ideal choice, but is not too horrible. Finally, every time a voter ranks a candidate in third place on a ballot, the candidate gets zero points for being the greatest of all evils. (This scheme is known as a *Borda count*, and, like instant runoff voting, it is often floated as an alternative to plurality votes.) Total up the points. All of a sudden, the election looks very different:

2,722,653 points for Norm Coleman
2,585,760 points for Al Franken
3,278,940 points for Dean Barkley

Dean Barkley wins handily—by virtue of being the least loathed candidate overall. Franken voters prefer Barkley to Coleman, and

Coleman voters prefer Barkley to Franken. Barkley is the compromise candidate that everybody can live with.

In the Minnesota Senate election, three different voting schemes—plurality, instant runoff, and Borda—would have yielded three different victors, even though the voters' preferences were identical in all three hypothetical scenarios. You can take exactly the same ballots and look at them in three different ways and come up with a perfectly valid argument about why each candidate should be elected to the Senate.

This illustrates a central problem with voting. Reasonable people can look at the same pile of ballots and come to very different conclusions about who should win an election. While plurality voting, Borda voting, and instant runoff voting each have their advantages and disadvantages,* none can claim to be the fairest way of electing a politician; they're all flawed. It's a mathematical truism known as *Arrow's theorem.*

In the 1950s, economist Kenneth Arrow proved that it's impossible to have a perfectly fair election system. But what does "fair" mean in this context? Well, there are certain characteristics that you would expect a fair election to have. One seems ridiculously obvious: there can't be a dictator who determines the outcome of an election; there can't be an individual whose vote overrules everybody else's. A fair election implicitly follows this "no dictators" rule. Another obvious condition: a vote shouldn't flout the unanimous will of the people. If everybody in the nation votes for Ross Perot, then, by God, Ross Perot had better win the election. A fair election also implicitly follows this

* As do innumerable other voting methods that election reformers like to talk about, such as Condorcet voting, approval voting . . . the list goes on and on. In my view, plurality voting is the best system because its ballots are simplest and therefore minimize voter error.

"unanimous vote" rule. Finally, there's a third characteristic that's a little less obvious: the ranks of candidates in a perfectly fair election should be a faithful reflection of the voters' preference. If, in a head-to-head race, Ronald Reagan would always beat Jimmy Carter, it's clear that the population genuinely prefers Reagan to Carter. Thus, in a truly fair election, Ronald Reagan should always come out ahead of Jimmy Carter when the results are tallied, no matter what other candidates are running. Under no circumstances should Jimmy Carter ever be able to beat Ronald Reagan; if he did, that would be an imperfect reflection of the voters' true preference. Thus a perfectly fair election should implicitly follow this "faithful reflection" rule as well.

Arrow's theorem proves that these three conditions of a perfectly fair election—"no dictators," "unanimous vote," and "faithful reflection"—are mutually contradictory. It's mathematically impossible to have all three at the same time. This means that there is no such thing as a perfectly fair election.

In practice, elections fail the "faithful reflection" condition. They don't give a simple tally of voters' preference of one candidate over another; they're much messier than that. Even though people preferred Ronald Reagan to Jimmy Carter, Carter theoretically could have won the election if an attractive third-party candidate also ran for president in 1980, splitting the Republican vote. If this happened, the "faithful reflection" condition would have been violated, because Carter would have won the election despite the populace's preference of Reagan to Carter. In other words, elections are so complex that no matter what voting system we use, the "wrong" person might wind up being elected. This is just a fact of life, something that we've come to live with in a democracy. No matter what method we use to elect our officials, there is some level of unfairness inherent in the process. It's inescapable.

Even though Arrow's theorem ensures that no election can be perfectly fair, the level of unfairness is pretty mild. We're not too disturbed by the idea of third-party candidates mucking up an election. Indeed, it's part of what makes elections so unpredictable—and so interesting. However, Arrow's theorem accounts for just a tiny fraction of the unfairness that plagues elections in the United States and around the world. There are much, much more worrisome problems with our electoral process—problems that are quite purposeful. Politicians are trying very hard to turn a mildly unfair voting system into something that's mind-bogglingly unfair. Make no mistake: there are people who are attempting to undermine the very mechanisms of democracy in order to ensure that their ideological allies get elected—regardless of the will of the people. And they've got a powerful weapon in the struggle: proofiness.

People have been undermining democracy by tampering with our electoral system for years. In the United States, the tradition is almost as old as the nation itself. By the early 1800s, scheming politicians had already created a powerful—and legal—method for staying in office even when the public tried to vote them out. This method is named after a gentleman whose signature is on the right edge of the Declaration of Independence, not far below the signatures of his fellow Bostonians Sam and John Adams. His name was Elbridge Gerry.

Gerry became governor of Massachusetts in 1810, but his party, the Democratic-Republican, was becoming increasingly unpopular. He and his allies were losing ground to their rivals, the Federalists, and they were terrified of what the upcoming election of 1812 would bring. If the voters had their way, the Democratic-Republicans would be kicked out of the statehouse, leaving the Federalists in power. From Gerry's point of view, this had to be avoided at all costs.

One of the powers of state government is to redistrict: to change the boundaries of voting precincts, altering which regions would be represented by which state senators. The Democratic-Republicans realized that if they got really creative with the way they redrew those boundaries, they could hold on to most of the seats in the state senate, even if the majority of voters in the state went Federalist.

When the Democratic-Republicans enacted the plan, it was extremely controversial. The bizarrely shaped new districts looked hideous and unnatural. A salamander-like district curled its body around Essex County, its belly to the west and its head snaking across the north. Wags promptly named it after its creator. The gerrymander was born.

As ugly as it was, the bizarre creature worked its magic for the Democratic-Republicans. Even though the Federalists got the majority of votes cast for state senator—50.8 percent of them, to be precise—the Democratic-Republicans won an overwhelming majority of the state senate seats: twenty-nine out of forty. By gerrymandering, they had turned what should have been an electoral defeat into a landslide victory. They held on to power not because of the will of the people, but in spite of it. They used gerrymandering to undermine the electoral process, annulling the votes of their opponents.

A gerrymander is a creature born from proofiness. At its core, gerrymandering is cherry-picking, with one key difference: the data

THE GERRY-MANDER.

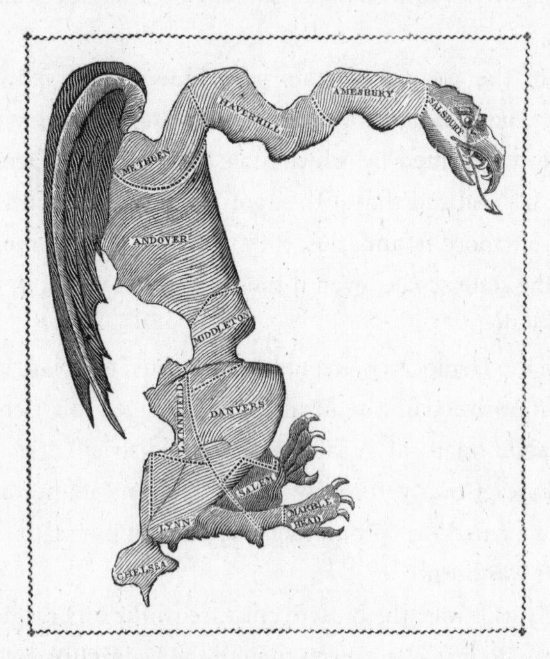

Figure 11. The original gerrymander.

being manipulated are votes. Indeed, the gerrymander is essentially a monster that allows politicians to carefully select votes, choosing those that they like and ignoring those that they don't.

Gerrymandering gets its power from two kinds of vote manipulation—two tricks that politicians have become extremely adept at over the years. These tricks are known as *packing* and *cracking*. Packing takes opposition votes and packs them tightly together, rendering most of them redundant. Cracking splits apart opposition strongholds, distributing their votes among multiple districts so that the enemy is not able to wield a majority in any one district.

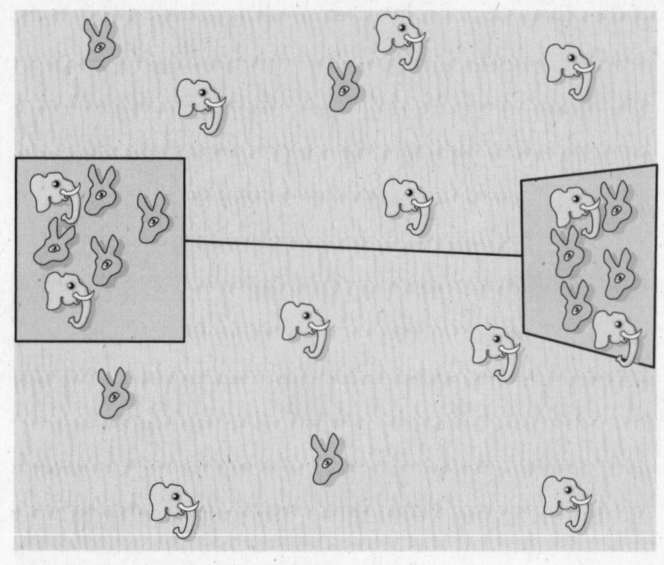

Figure 12. A fair division of districts.

As an example, imagine there's a county that is divided into four districts, two urban and two rural, each of which gets one representative to Congress. This county happens to be evenly divided between Democrats and Republicans; the two urban districts have a Democratic majority and the two rural districts have a Republican majority.

Each district sends a representative to the statehouse; the cities elect Democrats to represent them, while the two rural districts, naturally, elect Republicans. This is the way it should be; an evenly split electorate should split their representatives equally too. But imagine that the Republicans gain control of the statehouse, allowing them to redraw the boundaries of the districts. Their gerrymandering strategy will be to pack as many Democrats as they can into a single city district. They'll crack the other city apart, distributing

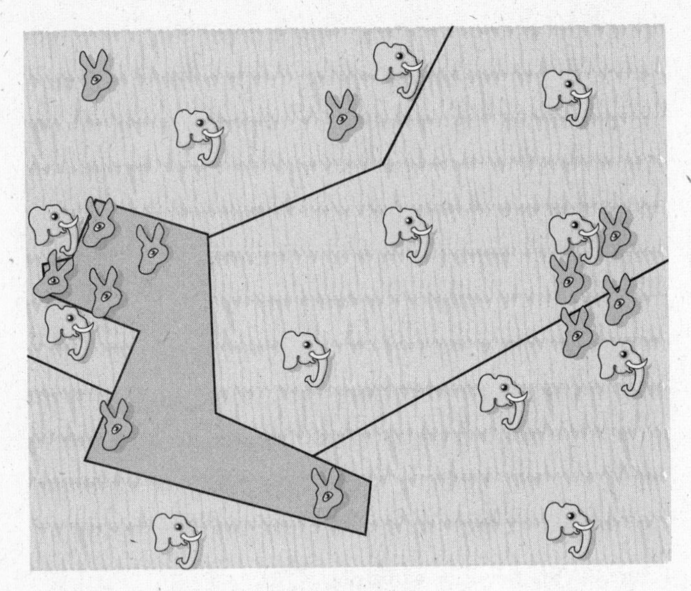

Figure 13. A pro-Republican gerrymander.

the city Democrats among Republican-heavy rural districts so that they're unable to muster a majority anywhere. The result is a bunch of districts that make a little less geographical sense—urban voters are mixed with rural voters—but gain the Republicans an extra seat in Congress.

Even though the electorate is split down the middle—there are exactly as many Democrats in the county as there are Republicans—the gerrymandering has allowed the Republicans to control 75 percent of the congressional seats. Conversely, if the Democrats had managed to gain control, they would do the exact same thing in reverse. They would pack Republicans into one rural district and crack the other, distributing Republicans among the three remaining districts so that they're firmly in the minority.

The Democratic seat-grab is just as effective as the Republican

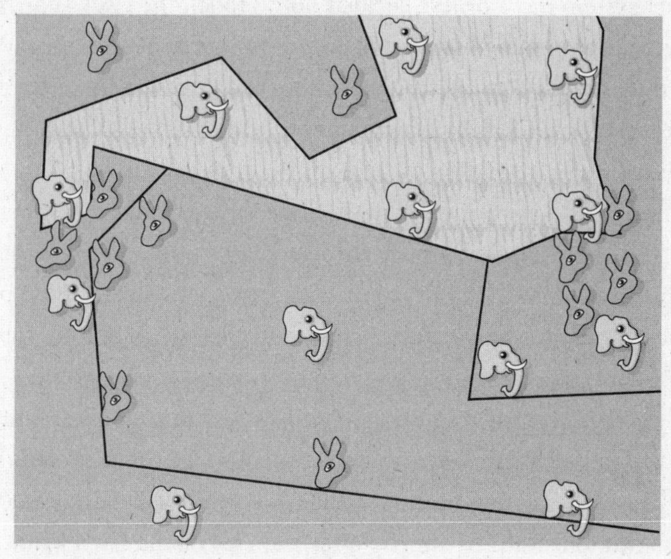

Figure 14. A pro-Democratic gerrymander.

version; they have 75 percent of the congressional representatives despite having only 50 percent of the vote.

In skillful hands, the gerrymander can give a party more power than the people want them to have; it can entrench an unpopular politician or dislodge a popular one; it can render some votes moot while investing others with great weight. Even worse, the practice of gerrymandering makes it more difficult for voters to punish the politicians who are robbing them of their votes—their redistricting plans make it extremely unlikely that the incumbents responsible for the gerrymander will lose the next election.

Gerrymandering is a direct affront to the democratic process; by allowing politicians to cherry-pick votes, it systematically undermines the validity of elections. But it's not easy to get rid of, because the issue is fiendishly complex. For many years, the federal court system more or less decided not to touch the issue at all, declaring it something that should be fought over by politicians rather than by judges. But the courts couldn't ignore it indefinitely, because gerrymandering was gnawing at fundamental constitutional rights.

The U.S. Constitution dictates that each state gets a number of seats in the House of Representatives proportional to the population of each state. As the population grows and shifts, the representation has to change, so states need to change the boundaries of their voting districts in response to new census numbers every decade. (These census numbers are also subject to proofiness—more on this shortly.) However, the power to change those boundaries isn't absolute. It's constrained by the Fourteenth Amendment to the Constitution. Ratified shortly after the end of the Civil War, the amendment guarantees that all citizens are given "equal protection" by the laws of the land. That is, each citizen is seen as equal in the eyes of the law—which in theory means that each person's vote should have the same value. By the 1960s, the courts were forced to recognize that gerrymandering, in some circumstances, was making certain citizens' votes count less than others'—some people's votes were being diluted. Judges couldn't continue ignoring the issue, thanks largely to the civil rights movement.

Shortly after the end of the Civil War, former slaves were suddenly full citizens—and like other citizens, they were entitled to vote and given the same protection under the law as any other citizen. That's the way it was on paper. In reality, though, former slave-owning states used all sorts of tricks, such as poll taxes, that were designed to keep African Americans away from the voting booths. (More on this later in the chapter.) They also used gerrymandering, cracking populations of African Americans so that they didn't get any representation in Congress. Despite constitutional guarantees, African Americans were being treated as second-class citizens.

In the mid-1960s, however, the civil rights movement changed the political landscape. Legislators and the courts finally attacked the problem head-on, outlawing the dirty tricks of racist politics. The Twenty-fourth Amendment to the Constitution outlawed poll taxes. A set of court decisions set the "one person, one vote" standard that dictates that each citizen's vote must be given roughly equal weight. The Voting Rights Act of 1965 outlawed procedures—including gerrymandering—intended to deny citizens their voting rights based on race or color. This made gerrymandering illegal, but only when it's done for racist motives.* Political gerrymandering—redistricting to gain political advantage—was still perfectly fine.

It would be simple except for the fact that race and politics can't be disentangled. In modern times, African-American voters are overwhelmingly Democratic. Latino voters also tend to support Democrats. So issues of race are always deeply political. Protecting the voting power of minorities is more or less tantamount to helping out the Democrats.† Making a distinction between racial gerrymandering and political gerrymandering is somewhat artificial. Even ignoring that fact, political gerrymandering is diluting citizens' votes just as surely as racial gerrymandering is.

Redistricting law, to put it politely, is a mess. The Supreme Court

---

* Confusingly, the Voting Rights Act had been interpreted as *encouraging* racial gerrymandering to give underrepresented minorities more power. However, Supreme Court cases in 1993 and 1995 functionally put an end to the practice.

† In the 1990s, Texas Republicans packed Democratic districts in a way that made a small number of African-American Democratic congressmen displace a larger number of white Democratic ones, so the gain for the African Americans was a loss for the Democratic Party. It was a fiendishly clever plan that its inventors dubbed "Project Ratfuck."

has been writhing in self-contradictory paroxysms trying to avoid addressing the issue. After years of pretending that political gerrymandering wasn't a topic suitable for lawsuits, in a 1986 decision the Court finally decided that gerrymandering for purely political purposes might theoretically fall afoul of Fourteenth Amendment protections. However, the Court gave no hints about what makes a partisan gerrymander cross the line into unconstitutionality. As a result, the ruling decided absolutely nothing; if anything, it made the matter more confusing. The first time the decision was tested in the Supreme Court—eighteen years later—the waters got even murkier. In a five-to-four party-line split, the Supreme Court declared a political gerrymander perfectly constitutional. Four of the majority justices went further. They declared that the Court was powerless to declare even the most obviously political gerrymanders unconstitutional, because there was no standard for determining whether a particular gerrymander was politically motivated or not. Anything goes.

It was the Wild West for political gerrymandering, and even the few gentlemanly rules that seemed to hold people's ambitions in check were dissolving. In 2001, in response to the new census numbers, a split Texas legislature finally compromised on a redistricting plan. It was a long and bitter fight, but it was over until the next census. Or so the Democrats thought. When the Republicans won both houses of the legislature in 2002, they re-redistricted, flouting the once-a-decade tradition. The Democrats attempted to stop the re-redistricting, fleeing the state so that a vote couldn't be called—while the Republicans called in the Department of Homeland Security to track down the wayward legislators—but the plan eventually went through, giving Republicans control of twenty-two of thirty-two legislative districts in the state. The case—or more pre-

Figure 15. District 12 in North Carolina, one of many gerrymander monsters.

cisely, four cases—went to the Supreme Court, which decided that except for one predominantly Latino district that had been illegally cracked apart in violation of the Voting Rights Act, the gerrymandering was hunky-dory. Even though the redistricting was not tied to the census and was only carried out for bald political gain, it was just fine.

At the moment, gerrymandering, so long as it's not racially motivated, is perfectly respectable in the United States. Though it allows politicians to cherry-pick votes, functionally allowing them to dilute the unfavorable ones, the courts don't seem inclined to correct the problem.* As a result, many of our voting districts remain

* Gerrymandering has a solution. Mathematicians have ways of spotting politically mandated gerrymandering, so it's possible to set standards. And some states have external, nonpartisan committees that are in charge of redrawing boundaries in response to changing populations.

so twisted and distorted that they put Elbridge Gerry's original monstrosity to shame.

Until there's a major change in the way the Supreme Court views the practice, gerrymandering proofiness is here to stay.

Gerrymandering is just one mathematical threat to democracy. There's another form of proofiness that's even more dangerous because it's less overt. This method has enabled politicians to deprive opponents of their votes. Even more than that, it renders them nonexistent—turns them into nonpeople who don't have the right to be represented in Congress. And the worst part is that the authors of this scheme are the very people who are supposed to be a last check against the excesses of the politicians in the government: this particular brand of proofiness comes directly from the Supreme Court of the United States.

The proofiness in question is a form of voter suppression—keeping "undesirable" votes away from the polling places. There's a long history of voter suppression in the United States, and as with gerrymandering, the political monkey business is inseparably mixed up with racial nastiness. After the Civil War, former slave-owning states used all sorts of tricks to prevent emancipated slaves from exercising their right to vote. They would create barriers to voter registration that were particularly burdensome to African Americans, who tended to be in the poorer and less well-educated segments of society. For example, in a number of states, people weren't allowed to vote unless they had paid a small fee—a "poll tax"—which of course hit poor African Americans much harder than richer citizens. (Especially since many white voters were exempted from paying the tax because of a "grandfather clause.")* As a result, the tax prevented many of them from voting.† The states pretended that the taxes had a legitimate purpose—they were intended to raise revenue, and they argued that people who paid state taxes became more interested in furthering the state's welfare. In reality, though, poll taxes were simply intended to keep African Americans from voting. Similarly, literacy tests were supposedly instituted to ensure that voters were able to make informed decisions; instead, they had the effect of barring less-educated African Americans from going to the polls. Legislators and judges eliminated these particular forms of voter suppression in the 1960s, but there are other more subtle forms that are still a problem. Voter ID laws, for example, are touted as a way to reduce voter fraud, as are periodic purges of voter registrations. However, there's very good reason to believe that these measures are being enacted because they have the indirect effect of reducing the number of African-American, Latino, and other minority voters.

However, none of the forms of voter suppression are more effective or more insidious than one engineered by the Supreme Court. In a series of decisions, the Court dressed a mathematical lie in the mantle of truth, wiping millions of people out of existence with the

---

* These sorts of clauses tended to exempt a person from the poll tax if he could prove that his grandfather had the right to vote—which white folk usually could and African Americans could not. Nowadays, shorn of its original racist heritage, a grandfather clause only refers to an exemption from a new law based upon prior circumstances.

† Interestingly, the "poll" in "poll tax" didn't specifically refer to voting, even though functionally it was a tax on going to the polls. The term comes from Middle English—*polle* meant "head," so a poll tax was in fact a tax put on each person's head (also known as a "capitation"). A polling place, on the other hand, is a place where your head is counted, so there's a shared etymology. Nevertheless, it's something of a coincidence that poll taxes were used to keep people from the polls.

stroke of a pen. Thanks to these rulings, more than 1 percent of the population of the United States consists of ghostlike disenfranchised creatures—citizens who in theory have the right to vote but are deliberately ignored. It's a stunning case of proofiness that goes right to the heart of what democracy is all about.

This particular scheme has to do with manipulating the U.S. Census. This may not seem like such a sinister plot, but at its root, a democracy is a government based on counting—on counting its citizens and their votes. The founding fathers of the United States recognized the importance of counting to their new government. Indeed, only five paragraphs into the U.S. Constitution, there's a passage that dictates that the government must perform an "actual enumeration" of its citizens every ten years. This decennial census is crucial to the functioning of the Republic, because it determines how much power different groups get to wield in the House of Representatives.

The 435 representatives in the House are divided (roughly) equally among the citizens of the United States—each representative nowadays votes by proxy for a block of roughly 700,000 people. The more citizens that a state has, the more representatives it gets, and the more power it wields in Congress. As the population shifts, political power (and money) follows. As the Northeast of the country atrophies, New York and Pennsylvania have been losing their preeminence to California and Texas. The political fortunes of a region—and of those who live in that region—hinge upon the results of the decennial census.

The government spends an ungodly amount of time and money to make an accurate count of its citizens; in 2000, the census cost roughly $6.5 billion—more than twenty dollars for each man, woman, and child in the United States. It's an incredible undertaking, and it's about as accurate as such a measurement can be. Unfortunately, the census, like any measurement, is fallible. And since the 1940s, statisticians have been forced to admit a depressing fact: no matter how hard census workers try, there's a systematic error that they can't get rid of. They can't count everybody.

In some ways, the census is like a monstrous government-run poll, but there's one very important difference. Instead of querying a sample of households and extending those results to the entire population, the census attempts to reach every single household in the United States; in theory, there's no extrapolation needed. So, just as in the case of voting, there's no statistical error. There isn't any worry that a statistical fluke makes the census sample look different from the entire population because the sample *is* the entire population. The census workers only have to worry about systematic errors. And there are quite a few to worry about.

Every poll relies upon the cooperation of its subjects—a poll can't record the opinions of people who toss their reply card in the trash or who slam down the phone when they hear the voice of a pollster. As a result, all polls are subject to "volunteer bias" that can inject an enormous amount of error into the poll. The census is no different. Every single household in the United States gets a census form, and the majority fill it out, but quite a few don't. In 2000, roughly one in three households didn't bother to return their questionnaire. To get an accurate count of the population, the Census Bureau still has to count the citizens in the households that refused to respond. This is where the big spending comes in. The bureau dispatches thousands of census workers who spend months going from household to household tracking down nonrespondents. The harder it is to get a household to respond, the more money is spent to try to contact the people in that household, but the bureau keeps trying until they run out of time and are required, by law, to give Congress the results. By the end of the process, the bureau manages to wring data out of all but about 2 percent of the population. It's an extraordinary effort. But it still is full of errors.

Not only has the census failed to reach 2 percent of the population; it accidentally double-counts about 1 percent. This means that for all that effort, the census is only good to within about ten million people, plus or minus. This plus or minus is enormously important, politically; these ten million people would be entitled to roughly fourteen representatives in the House. It's incredibly disheartening; all that time and money spent, and errors in the census are still huge. These errors are impossible to correct by ordinary means. The government could theoretically stake out the homes of every single nonrespondent, but that would cost astronomical amounts of money, and even this wouldn't manage to catch everybody. Even with double its current budget, the Census Bureau can't do much better with its measurements than they already are. However, the situation isn't hopeless. There is a way to reduce these errors enormously by using a set of statistical tricks known collectively as *sampling*.

The best way to understand sampling is through an example. Imagine that there's a shallow pond that's full of trout and minnows. The government has hired you to count how many fish the pond contains. You row gently from one end of the pond to the other, counting the fish that you see along the way. You come up with a count of 599 trout and 301 minnows. Your grand total is 900 fish in the pond, about 67 percent of which are trout and 33 percent minnows.

As you can probably guess, the answer is off because your count is error-prone. One source of error is that the fish are constantly moving about, making it all but certain that you'll count some fish twice and others not at all. Another source of error is that minnows are harder to spot than trout. They're tiny and timid; they tend to hide when the boat comes nearby. So it's quite likely that you're undercounting minnows—and no matter how many times you count from your boat, minnows are likely to be underrepresented. Conversely, big, visible trout are more likely to be double-counted.

You can correct for these errors, but to do so, you have to make another measurement to figure out how bad they really are. After you've done your initial survey of aquatic life, you do another boat count of a small, representative section of the pond and record the numbers of fish that you find (say, 30 trout and 15 minnows). Then you make a more careful (and more invasive) count of that small section. Net off that little region of the pond, dredge up every single fish in that area, and pull them into the boat. Counting them as you toss them one by one back into the pond yields an incredibly accurate count: say you find that there are really 28 trout and 19 minnows.

This new information tells you how accurate your boat count really was. The data tell you that you did in fact overcount trout (you counted 30 from the boat, but there were really 28) and undercounted minnows (you counted 15 from the boat, but there were really 19). And now that the data tell you the nature of your measurement errors, you can correct for them. You now know that your original count of 599 trout is too large and should be adjusted downward—to about 560—to compensate for your tendency to overcount trout. Similarly, your count of 301 minnows is too small and should be adjusted upward—to about 380—to account for timid minnows that you were unable to see from the boat. Your new, adjusted total is 940 fish in the pond, about 60 percent of which are trout and 40 percent of which are minnows.

The new numbers aren't perfect by any means. It's possible that the small netted-off section of the pond was not truly representative of the entire pond. There might have been a particularly dense and hard-to-spot concentration of minnows in the area, for example.* Also, since you're extending your observations about a small number of fish to the entire pond, you have to worry about statistical errors that would be irrelevant in a direct count of the entire population. However, the increase in statistical error is more than compensated for by the decrease in systematic error—your measurement allows a dramatic reduction in the problems caused by miscounting certain segments of the population. In short, you're trading large, systematic errors for (hopefully) smaller, mostly statistical errors—and the result is a better, more accurate count.

This is sampling in a nutshell. By looking extremely carefully at a sample of the population, the Census Bureau can generate data that allow it to correct for the systematic undercounts and overcounts in the census. From a statistician's point of view, it's a no-brainer. A corrected count would produce a much more accurate depiction of the population of the United States than a count-every-head census ever could. Instead of having censuses that are good to within a few percent, it would be possible to reduce the errors down to a fraction of a percent. The most accurate tally of the population of the United States would not come from a straight head count; instead, it should be a census that is corrected by sampling. As an added bonus, a census that uses sampling is cheaper than a straight head count. Instead of spending billions of dollars to try to chase down that recalcitrant last few percent who don't respond to census

* Due, no doubt, to the presence of that most feared of aquatic creatures, the statistical fluke.

workers, the bureau can spend a few tens of millions doing the same thing, even more exhaustively, in a small number of communities and use that data to correct for the undercount. Sampling is more accurate and it's cheaper. So every politician should be in favor of it, right?

Not quite. Unfortunately, sampling is caught up in the racial politics of voter suppression. The citizens who tend to be undercounted by the census tend to be poorer people who rent their homes rather than own them. A disproportionate number don't speak English and are distrustful of government authorities (including the Census Bureau). They tend to be minorities—and they tend to vote Democratic. Conversely, the overcounted tend to be white and affluent, and are more likely than not to vote Republican. If the United States were a pond, minorities would be the minnows while whites would be the trout. The moment you use sampling to correct for the undercount, you suddenly add several million more minorities—Democrats—into your count of the population. It's something that Republicans want to prevent so badly that they are forced to take an idiotic stance: they insist the proper way to conduct a census is the least accurate and most expensive method.*

The Census Bureau was reduced to reporting two population numbers to Congress every decade: a sampling-corrected number that statisticians and population experts use because they need precise data to estimate everyday population trends, such as poverty

---

* The opposition to sampling, as with other forms of voter suppression, doesn't run ideologically deep in the Republican Party any more than pro-minority, pro-voting-rights sentiments run deep in the thoughts of mainstream Democrats. If the roles were reversed—if it were primarily Republicans who were being undercounted—there's little doubt that Democrats would be trying to suppress sampling while Republicans would be championing it. It's all petty-minded scrabbling to gain a political advantage.

rates, incomes, and household sizes; and the highly error-ridden head count that Congress insisted on using to reapportion House seats. When in the late 1990s the Census Bureau finally proposed presenting only its best, sampling-corrected number to Congress, Republicans in the House of Representatives promptly sued. Using sampling to correct the numbers, the Republicans argued, was unconstitutional; the "actual enumeration" required by the Constitution had to be a simple head count unspoiled by any statistical mumbo-jumbo that might make it more accurate.

The case went all the way up to the Supreme Court. The American Statistical Association filed a brief with the court that made the case pretty clear: "Properly designed sampling is often a better and more accurate method of gaining such knowledge than an inevitably incomplete attempt to survey all members of such a population. . . . There are no sound scientific grounds for rejecting all use of statistical sampling in the 2000 census." But the Supreme Court disagreed. In a five-to-four decision—the five most conservative judges versus the four most liberal—the Court determined that sampling was illegal. The apportionment of House seats, by law, had to be based upon flawed, highly error-prone population numbers that undercounted minority voters.

Even though the ruling evaded the question about whether the use of statistics ran contrary to the Constitution—whether the "actual enumeration" clause referred to a head count and nothing else—there's no question that the conservative majority was hostile to the whole concept of sampling. In a concurring opinion penned by Antonin Scalia, one of the most conservative justices of the Court, Scalia strongly implied that any use of statistical techniques would make the founding fathers spin in their graves.

Scalia turned to eighteenth-century dictionaries to show that the phrase "actual enumeration" had to mean counting each individual person, one by one; he cited the 1773 Samuel Johnson dictionary, for example, which defined "enumerate" as "To reckon up singly; to count over distinctly; to number." Aha! To reckon up singly! Scalia pounced: by using the term "enumerate," the founding fathers meant to count each person, one by one. Thus the census "requires an actual counting, and not just an estimation of number." Further, using statistical techniques will "give the party controlling Congress the power to distort representation in its own favor." Only head counts, as inaccurate as they are, are free from manipulation. Thus sampling is unconstitutional.

This is a specious argument on several counts. First, even if Johnson's dictionary was the key to the founding fathers' intent, the definition "to number" is just as valid as "to reckon singly." Besides, the word "enumeration" was merely an accident. It was inserted during the drafting of the U.S. Constitution by the Committee on Style, a group that made minor, nonmeaningful changes to the document to correct grammar and clarity. They changed the word "census" to "enumeration" for reasons unknown.* So to dwell upon the finer points of the dictionary definition of "enumeration" is ignoring the fact that the founding fathers called for a "census," and that "enumeration" was substituted for reasons of style.

More important, there's no bright-line distinction between "actual counting" and "estimation." As we've seen, counting is a measurement like any other, and is thus subject to error. This error is unavoidable; it turns even the best count into nothing more than an

* One theory is that people were afraid of taking a census that wasn't approved by God. According to the Bible, when King David called for an unsanctioned census, God punished him by sending a pestilence that killed 70,000 Israelites in three days.

estimate, an approximation of the truth. And if there are other measurement techniques that give you a better approximation of the truth—such as statistical sampling—they deserve the title of "actual" more than counting does. Pretending otherwise is to place too much faith in the error-prone numbers that come from a head count: it is an act of disestimation.

Finally, the idea that head counts are free from manipulation is wrong. In fact, even the purest, most pristine head count that the U.S. Census can possibly perform is subject to statistical tinkering. It has to be. Census Bureau workers have to be able to interpret and even alter the data using statistical tools, otherwise the census would be utterly meaningless.

Imagine, for example, that some joker in Sitka, Alaska, fills out his census form to say that there are 300 million people living in his household. If the Census Bureau were to take him seriously, it would mean that Alaska would suddenly be the most populous state in the Union by a huge margin; indeed, half of the representatives in the House would be representing this gentleman's household. Luckily, no census worker is stupid enough to believe him. It's obvious that the guy is lying—he gave the census a bad piece of data. But what can the Census Bureau do about it? The only choice is to clean up the datum somehow—and doing this means that they must use a statistical technique known as *imputation*.

In an imputation, a Census Bureau statistician picks out a datum that looks wrong. (Anyone who says that he has seventy-seven children or is 175 years old, for example, is probably lying.) Then the statistician wipes out the questionable answer and replaces it with census data from similar-looking households. The replacement number is a guess, but an educated one—and it's certainly closer to the truth than the phony datum. And in fact, there's really no alternative. Wiping out the datum or, more drastically, tossing out the entire census form is also imputation. The act of wiping out a datum is a substitution: the worker is still replacing someone's answer (seventy-seven children) with another answer (zero children); a null answer is still an answer. Similarly, tossing out a census form is equivalent to imputing that a dwelling is vacant. Instead of making bad imputations by simply wiping out dubious results, the Census Bureau prefers to make an educated guess from the freshest census data it has, a process known as "hot-deck imputation."* It's more likely to be approximately correct, so it does less violence to the validity of the census results. The only other option—the only way to avoid imputation entirely—is to take every single census form at face value. You have to duly record the responses of every 175-year-old woman, every man with seventy-seven children, and, yes, the gentleman in Sitka who has 300 million people in his household. Without imputation, the results of the census become worthless.

The Supreme Court decision about sampling was effectively a ban on using statistical mumbo-jumbo, but imputation is a form of statistical mumbo-jumbo that wasn't addressed by the previous decision. So when the (sampling-free!) 2000 census results were released, the state of Utah, which was denied an extra representative in Congress, sued. They argued that the bureau's use of imputation was illegal, and in 2002 this case worked its way up to the Supreme Court.

Utah's case put the court in a bind. If the justices ruled that

* The "deck" was a deck of punch cards, which shows how long this technique has been in use—since the first half of the twentieth century.

imputation was unconstitutional, it would have rendered the Census Bureau powerless to correct spurious data; the one joker from Sitka could theoretically render the entire count meaningless. However, if the Court decided that imputation was permissible, it had to split hairs to explain why one statistical technique—imputation—was acceptable while another—sampling—was illegal.

In another five-to-four decision—the liberals were joined by the usually conservative chief justice, William Rehnquist—the Court decided to take the latter course. In a shining example of how many justices it takes to split a hair, the Census Bureau was allowed to continue using imputation, at least for the moment. However, the minority lobbed grenades at the decision, accusing the bureau of using illegal, and perhaps unconstitutional, statistical witchcraft. Sandra Day O'Connor wrote that imputation was simply a form of sampling and should thus be banned. Clarence Thomas essentially repeated Scalia's argument from the earlier sampling case (even recycling the dictionary definitions of "enumeration") but turned up the volume a little bit. He insisted that the founding fathers were "well familiar with methods of estimation," so had explicitly rejected the sophisticated techniques used by the Census Bureau. (Which, as those techniques were developed in the twentieth century, would be quite a feat.) And Thomas repeated the canard that "actual counting" is fundamentally different from "estimation" even though counting *is* an estimate. Once again, the justices marshaled proofiness to justify using bad numbers instead of good ones—and to try to ensure that certain people are robbed of representation by virtue of a purposefully inaccurate census.

As this book goes to press, the Census Bureau is beginning its 2010 census. The Supreme Court's two rulings leave census law in a complete shambles; there's a sense that some statistical techniques are kosher while others are illegal, and there's no real basis for telling which are which. Republican legislators are already challenging the validity of the new census. They threatened to sue long before the first census form went out in the mail, and a few have gone beyond mere threats. In October 2009, Louisiana senator David Vitter—probably best known for his use of a high-end D.C. escort agency—tried to force the Census Bureau to rewrite its census forms to ask respondents to declare whether or not they were U.S. citizens. If his measure had passed, not only would it have cost enormous amounts of money (425 million forms would have had to be thrown out) and delayed the start of the census, but it would also certainly have scared noncitizens, particularly illegal immigrants, from participating in the census.* It was a shameless play to try to make the undercount even worse. Predictably, it was defeated by a strict party-line vote, with the Democrats opposing and the Republicans backing it.

However, open season will really begin once the results of the new census are in. It's almost certain that the new decade will bring a fresh cluster of lawsuits about census methods, and, given the conservative makeup of the Supreme Court, it's quite possible that the first census of the twenty-first century will be forced to divest itself of all mathematical techniques that were developed after the eighteenth.

Proofiness has undermined the very foundations of our democracy—the mechanisms that we use to count our citizens and ensure that they are justly represented in the Republic. Gerryman-

---

* The Constitution requires that *all* persons be counted, not just U.S. citizens or legal immigrants.

dering for political gain is deemed acceptable, even though it clearly dilutes the votes of some of our citizens. Statistical sampling is deemed unacceptable, even though rejecting it forces the government to use numbers that it knows are inaccurate. No matter how many intellectual backflips legislators and judges go through to justify their positions, the fact remains: bad mathematics is being used to deny our citizens—mostly our minorities—their rightful vote. In a democracy, there can be no graver sin.

7

Alternate Realities

I believe that there is such a thing as objective truth, but a lot of readers have an objective truth that differs from mine.

—Cynthia Tucker, letters editor of the *Atlanta Journal-Constitution*

Governments can't change reality. But it's not for lack of trying.

Politicians will challenge—and try to alter—any fact that they dislike, no matter how incontrovertible it might be. At one point, Indiana even tried to change the value of pi.

Yes, that's the pi you learned about in math class—the ratio of a circle's circumference to its diameter. Apparently, Indiana state senator Taylor I. Record, egged on by an eccentric doctor, decided that pi's value (approximately equal to 3.14159) was inconvenient. It made calculations difficult. Therefore, it had to be changed. On January 18, 1897, House Bill number 246 was born—if passed by the House and the Senate, and signed by the governor, it would officially change the value of pi to 3.2. The old value, "wholly wanting and misleading," would be discarded for the greater glory of mankind.

The bill got off to a great start. The committee on education gave it the green light and a few weeks later the bill passed the House unanimously, sixty-seven votes to zero. Next up was the Senate. The bill was handed, appropriately enough, to the committee on temperance, which gave an enthusiastic thumbs-up. It would have passed the Senate but for the intervention of the chair of Purdue University's math department, who explained to the senators why passage of the bill would make Indiana a laughingstock. After a brief debate, in which one senator exclaimed that one "might as well try to legislate water to run up hill as to establish mathematical truth by law," the bill was shelved indefinitely. It was a rare legislative victory for the forces of mathematical reality.

Indiana's abortive attempt to rewrite the laws of nature is absurd, but it's not an isolated incident. Our governments misuse mathematics in subtler ways all the time, trying to banish facts that are embarrassing and inconvenient. And the most mind-bending denials of mathematical reality come from the branch of government that's supposed to be the guardian of truth: the judiciary.

Our courts have been infected with proofiness. Mathematical and statistical knowledge can be used to free the wrongfully convicted, to help convict the guilty, and to reduce bias and injustice by law enforcement and the courts. However, attorneys use proofiness to free the guilty; prosecutors use it to convict the innocent. All through the court system, bogus mathematical arguments are used to justify injustice. The problem goes all the way to the top; even Supreme Court justices use phony statistics to push their own political agendas. These lies go far beyond mere political tinkering, far beyond messing with votes and manipulating the census. They are used to try to distort the nature of justice—and of truth.

In the hands of the courts, proofiness is a weapon of terrifying power. The alternate realities that judges construct quite literally have determined the difference between life and death.

The justice system can't be totally free of lies and distortions; after all, courts are chock-full of lawyers. For many of them, glibness is a virtue—for a defense attorney, a specious argument might be just the thing to get a client off. And there's nothing that's better for confusing a jury than proofiness.

The famed defense attorney Alan Dershowitz has a well-earned reputation for running circles around prosecutors. His finest moment came in 1995 when he and his team got O. J. Simpson acquitted of a double murder. Despite an extraordinary amount of forensic evidence that seemed to prove Simpson guilty, Dershowitz and his colleagues threw up a cloud of obfuscation that allowed his client to walk away from the charges. And proofiness was a key element of that cloud.

Simpson had been arrested a number of years earlier for battering his then wife, so when she was stabbed to death, it was natural to consider Simpson a key suspect. Dershowitz, however, turned that piece of evidence completely upside down thanks to some phony probabilities. He convinced the jury that the battery made it incredibly improbable that Simpson murdered his ex-wife; after all, only one in a thousand wife-beaters winds up murdering his spouse. One in a thousand! Such a small probability means that O. J. Simpson almost certainly isn't the murderer, right?

It's hard to express just how wrong this argument is. It's tantamount to turning Simpson's wife-beating—a powerful indication of violent tendencies toward his spouse—into exculpatory evidence. And in fact, Dershowitz's line of reasoning is transparently fallacious; it fails to take into account that the probability of being murdered is very small. A one in a thousand chance of a woman's being killed by an abusive spouse is huge by comparison. When you crunch the probabilities properly, it becomes clear that if an abused woman is murdered, it is highly likely that the murderer is her abuser. (Using reasonable assumptions, separate groups of statisticians calculated that the probability is more than 50 percent—perhaps as high as 80 percent.) Dershowitz's argument was nothing more than a ballsy case of risk mismanagement; with a quick shuffle-step, he misled the jury about the probability that Simpson was a murderer and turned a damning bit of evidence into an asset. The jury was apparently fooled. Because of this argument (and other nonmathematical ones) Simpson was acquitted.

Even though it was a bogus argument that helped put a likely murderer back on the streets, Dershowitz was doing his job. Defense attorneys are supposed to use every legal resource at their disposal to protect their clients. They are the last line of defense when the government tries to rob someone of his freedom, and the price of a robust legal defense system is, on occasion, letting some criminals go free. Defense attorneys sometimes use proofiness to get their clients off, but it's not nearly so grave a sin as using proofiness to put someone behind bars.

There's a saying: any prosecutor can convict a guilty man, but it takes a great prosecutor to convict an innocent man. Proofiness is a wonderful tool for creating great prosecutors. By using an expert witness who spouts mathematical nonsense under oath, a prosecutor can make even a blameless person seem guilty.

One notorious case comes from the United Kingdom. Sally Clark, a British lawyer, was convicted in 1998 of murdering her two infant children. She was convicted in part because of the spectacular testimony of a famous pediatrician, Sir Roy Meadow, who used an incorrect probability to "prove" that Clark's children weren't victims of sudden infant death syndrome (SIDS). With SIDS out of the way, Clark had to be the killer.

Meadow made his name in the 1970s by identifying a syndrome that he named "Munchausen Syndrome by Proxy," a mental disorder where a parent (usually the mother) gains attention by faking—or causing—a child's illness. His career soon became a crusade to protect infants from their mentally ill parents. He was particularly concerned that many infants with the poorly defined catchall diagnosis of SIDS had in fact been murdered. He was so skeptical of the condition that at one point he suggested abandoning the term SIDS altogether.

In 1996, Clark's newborn son died, and the coroner treated it as a case of SIDS. But when her second newborn son died two years later, the coroner changed his mind. Within a month, Clark was accused of serial murder: smothering her first child and shaking the second to death.

Meadow was a key witness for the prosecution, and helped seal the case with the stunning statistic that the likelihood of having two SIDS cases in the same family was about 1 in 73 million—or as Sir Roy put it, it was more likely that a gambler would win four successive 80-to-1 bets at the racetrack than for Ms. Clark's family to suffer two cases of SIDS. A two-SIDS family, he argued, would only occur in Britain about once in a century. The jury found Clark guilty, and the judge sentenced her to life in prison.

The problem was that Meadow's testimony was worthless; it was a piece of risk mismanagement. The 1 in 73 million figure came from multiplying the likelihood of SIDS in an older, nonsmoking, wealthy woman's household (1 in 8,543) by itself. Unfortunately, this figure was vastly inaccurate for a number of reasons. SIDS seems to have a genetic component (as well as an environmental one), thus a family that experiences one SIDS death is at increased risk—the odds are rather greater than 1 in 8,543—for a second, so the overall odds are significantly greater than 1 in 73 million. Even more serious is that the figure is misleading in a similar way to the O. J. Simpson abuse statistic. In isolation, the likelihood of two SIDS deaths looks very small, but the proper statistical procedure needs to compare that probability to the probability of a person's being a serial murderer, which is even smaller. Just as it's better than even odds that a previously abused woman who winds up murdered has been murdered by her abuser, it's better than even odds that Clark's second child died of SIDS. Meadow's astronomical-seeming probability was greatly overstated because it failed to put that probability in the proper context. This particular variety of risk mismanagement is so common in courts that it's become known as the *prosecutor's fallacy*. (See appendix C for a fuller explanation of the prosecutor's fallacy in the Sally Clark and the O. J. Simpson cases.)

Luckily, a few sharp-eyed statisticians cried foul when they heard about Meadow's testimony. A blistering article in the *British Medical Journal* entitled "Conviction by Mathematical Error?" attacked Sir Roy's numbers, and when Clark lost her appeal, the president of the Royal Statistical Society wrote to the lord chancellor—then the head of Britain's judiciary—declaring the 1 in 73 million figure "invalid."

In retrospect, it appears that Clark was truly innocent. One fact that prosecutors never presented at her trial is that her second child had been suffering from a significant staph infection of the central nervous system—it was quite likely the cause of his death. Clark's second appeal in 2003 was successful, and three other women who had been convicted with the aid of Meadow's testimony were also set free.

Prosecutors rely on expert witnesses such as Roy Meadow to feed facts—real or fabricated—to juries to help their cases. The right expert can make it easy to clear even high legal barriers. In Texas, a convict can get the death penalty only if a jury finds that he would continue to commit violent acts that would be a threat to society. This should be a relatively hard thing to prove—unless you've got expert witnesses like psychiatrist James Grigson in your corner. Grigson was nicknamed "Dr. Death" because he waltzed into courtrooms to declare that a convict was a sociopath and would almost certainly kill again, clearing the way for a death sentence. He did so over and over—even in cases where the wrong person was convicted.*

The case of murderer Thomas Barefoot is an example of how Grigson used proofiness to snow juries. In Barefoot's case, Grigson didn't even give him the courtesy of a psychiatric examination before testifying, multiple times, that he was 100 percent certain Barefoot would kill again. It didn't matter whether Barefoot would be in jail or whether he would be free—Grigson asserted with perfect certitude that Barefoot would kill again. It's a stunning piece of risk mismanagement; there's no way that Grigson could be so certain. However, Grigson's cocksure testimony tended to impress juries—

* Errol Morris's documentary *The Thin Blue Line* gives a gripping account of one such case.

and it did in this case too. Barefoot was given the death penalty. The case went up to the Supreme Court.

The American Psychiatric Association wrote a brief that revealed Grigson's handiwork to be gibberish:

> The large body of research in this area indicates that, even under the best of conditions, psychiatric predictions of long-term future dangerousness are wrong in at least two out of every three cases. . . .
>
> To the extent such predictions have any validity, they can only be made on the basis of essentially actuarial data to which psychiatrists, *qua* psychiatrists, can bring no special interpretative skills. On the other hand, the use of psychiatric testimony on this issue causes serious prejudice to the defendant. By dressing up the actuarial data with an "expert" opinion, the psychiatrist's testimony is likely to receive undue weight. In addition, it permits the jury to avoid the difficult actuarial questions by seeking refuge in a medical diagnosis that provides a false aura of certainty.

In other words, Grigson's predictions were worse than worthless because they were dubious opinions dressed up in the clothing of scientific and mathematical certainty. At best, Grigson was guilty of risk mismanagement, as he misrepresented the probability of a convict's killing again. More likely, though, his predictions were complete nonsense with no rational basis; they were Potemkin numbers.

Three Supreme Court justices were appalled by Grigson's proofiness:

> A layman with access to relevant statistics can do at least as well, and possibly better; psychiatric training is not relevant to the factors that validly can be employed to make such predictions, and psychiatrists consistently err on the side of overpredicting violence. . . . Despite their claims that they were able to predict Barefoot's future behavior "within reasonable psychiatric certainty," or to a "one hundred percent and absolute" certainty, there was, in fact, no more than a one in three chance that they were correct.

But unfortunately for Barefoot, the other justices rejected the American Psychiatric Association's argument: "The suggestion that no psychiatrist's testimony may be presented with respect to a defendant's future dangerousness is somewhat like asking us to disinvent the wheel," the justices in the majority wrote. Using phony data was so important to the prosecutorial process that it couldn't be dispensed with. Besides, the data were acceptable, the justices wrote, because psychiatrists weren't "always wrong with respect to future dangerousness, only most of the time."

Four months later, shortly after midnight on October 30, 1984, Thomas Barefoot said, "Sharon, tell all my friends goodbye. You know who they are: Charles Bass, David Powell . . ." Barefoot coughed one last time and then went silent. He was pronounced dead—executed by lethal injection—at 12:24 a.m.

Barefoot was a loathsome character with a long rap sheet. He had shot a police officer while on the run; he was under suspicion of raping a three-year-old girl in New Mexico. He may well have deserved the penalty he received, but if he did, prosecutors shouldn't have had to rely on Grigson's proofiness to impose it. Yet they did, and not just with Barefoot. Prosecutors regularly relied on Grigson to dispatch their death penalty cases, using him to construct alternate realities where every person convicted of murder was a latent serial murderer who was just itching to kill again. For more than three decades, Grigson served as an expert witness at hundreds* of capital murder trials. In most cases, he would declare that the convict was a sociopath who deserved death—sometimes without even bothering to examine the patient. Grigson was well rewarded for his efforts. By the mid-1980s, he was reportedly earning $100 an hour for his death penalty testimony. In 1995, the American Psychiatric Association and the Texas Society of Psychiatric Physicians expelled Grigson for ethics violations, but prosecutors still found him indispensable. They continued to use him in the courtroom—where unethical conduct is apparently no disgrace—to send people to the death house. Grigson continued testifying in court until he retired in 2003.

There are many others like Meadow and Grigson out there, others who use proofiness to convict people or to send them to death. When they dress up nonsense in the garb of absolute truth—through the prosecutor's fallacy or through expert witnesses or other means—they undermine our justice system little by little.

When it comes to mathematics, it's chaos in the courtroom. Anything goes; any mathematical argument, no matter how specious, has a

* It's hard to pin down exactly how many. Grigson himself testified in the mid-1990s that he was a witness in nearly four hundred capital cases; other sources put the number at around one hundred fifty to two hundred. But by all accounts, in the majority of these he found the defendant to be a sociopath.

chance of carrying the day if the judge isn't alert enough. If the mood strikes, a judge might deploy his own proofiness to make a point.

A judge's ruling in an inconsequential case in the D.C. district court a few years ago is a good example. The case had to do with a disused rail line that went through two small cities in Missouri. The Surface Transportation Board—the body responsible for regulating railroads—had given permission for the railway line to start running two trains a day on the line. However, citizens of the cities sued because the railway failed to assess the environmental impact of the trains, as (seemingly) required when there is "an increase in traffic of at least 100 percent" on a rail line. This was the signal for the judge to go to town:

> The cities believe that an increase from zero tonnage to whatever gross tonnage is represented by 520 trains per year (10 per week) equals an "increase in rail traffic of at least 100 percent." How the cities calculate this is a mystery. The regulation asks the question: what is the percentage increase on the acquired line? Suppose there were 100 tons per year before the acquisition and 200 tons afterwards. One does not have to be a Richard Feynman to figure out that 200 tons is 100% greater than 100 tons. The formula 100 x (a / b) yields the percentage, when a equals the post-acquisition increase in tonnage (100 tons) and b equals the pre-acquisition tonnage (100 tons). But there is trouble when b equals zero, as it does here. Then there must be division by zero. Yet as mathematicians know, "you can't legitimately divide by 0. ∞ doesn't mean anything."

The judge insisted that the cities were dividing by zero, and therefore they were mistaken in asking for an environmental impact statement.*

For all of his paroxysms about logical paradoxes, the judge was constructing an alternate reality where a mathematical problem was insoluble, when in fact the answer is very simple. All you have to do is ask yourself three questions. First: is there an increase in rail traffic on the line? (Yes. It's going from zero to two trains per day.) Second: what would be a 100 percent increase over the current traffic on the line? (It would be zero trains: 100 percent times zero is zero.) The answers to these two questions allow you to answer the third, crucial question: is the increase in rail traffic on the line greater than or equal to a 100 percent increase over current traffic? (Yes. Two is greater than zero.) Since the answer to question three is yes, the regulation is satisfied—there is indeed an increase in rail traffic of at least 100 percent, so yes, an environmental impact statement is necessary.† No mind-bending logic games required.

Courtrooms are terrible when it comes to matters of mathematical or scientific truth; too often, the alternate reality of proofiness prevails over genuine reality. Charlie Chaplin found this out the hard way in 1944 when a mentally unstable young woman, Joan Berry, accused him of fathering her daughter. It looked bad for Chaplin—he admitted having sex with her, though he claimed that the affair had ended months before the child was conceived. In ad-

---

* This case came to my attention because my book *Zero*, as well as another book about the number zero, was cited as support for this opinion.

† Despite the numerical absurdity in the decision, I think the case was decided correctly. There's a line in the regulations that specifically deals with reinstituting service over a previously abandoned railroad, which seems to make it unnecessary to file an environmental report unless there's an increase of at least eight trains a day. (But the regulations are so badly written that it's hard to tell for sure.)

dition to the paternity suit, Chaplin was facing federal charges. Since the alleged liaison took place in a different state, the government went after him for violating the Mann Act—transporting a woman across state lines for "immoral purposes." (The prosecution was probably motivated, at least in part, by the fact that the FBI had labeled Chaplin as a communist sympathizer.)

When the trial began, Chaplin had a wonderful stroke of luck. Doctors tested his blood, along with the mother's and daughter's, and proved beyond any reasonable doubt that he was not the father. The child's blood type—type B—could not have been produced by a union between Chaplin (who had type O blood) and Berry (who had type A blood). As the doctors testified, "The result of these blood grouping tests is that in accordance with the well-accepted laws of heredity, the man, Charles Chaplin, cannot be the father of the child, Carol Ann Berry." Unless there was some mistake with the test results (which could easily be repeated if there was any doubt of misconduct), it was almost 100 percent certain that an unknown man, who had either type B or type AB blood, was the father. Chaplin was clearly innocent of the accusations.

It didn't matter. The prosecutor—and the courts—rejected the notion that the blood test was conclusive and pressed on with the trials, using such scientific evidence as a "resemblance" test, where jurors gazed intently at the baby to determine whether she looked like Chaplin. The Mann Act charges at first ended in a mistrial and then in an acquittal; however, in his paternity trial, Chaplin was found to be the father. The court's truth—the legal truth—was that Chaplin was the father of Joan Berry's baby, even though the Truth, with nearly absolute mathematical certainty, was otherwise. And the prosecutors and the judges knew it.

A truly just court system would treat as sacred any evidence that can show, with extremely high probability, that an accused person is innocent. Ours is not that system. In 2009, for example, the Supreme Court ruled, five to four, that William Osborne, a rape convict, had no right to get a high-quality DNA test on evidence that might acquit him. As the dissenting justices put it, "The DNA test Osborne seeks is a simple one, its cost modest, and its results uniquely precise. Yet for reasons the State has been unable or unwilling to articulate, it refuses to allow Osborne to test the evidence at his own expense and to thereby ascertain the truth once and for all." Nevertheless, his claim was denied.

Even when a condemned man's life hangs in the balance, the courts are willing to shut their ears to evidence that shows his innocence. In 1993, the Supreme Court ruled that Leonel Herrera, who had been convicted of killing two police officers, was not entitled to a new trial; he was barred from presenting new and fairly compelling testimony—including a note from the lawyer of the person who probably killed the officers—that might demonstrate his innocence. Herrera was executed by lethal injection four months later.

Unfortunately, there's no place where proofiness is more apparent than death penalty cases. The machinery of death is oiled with falsehoods.

In the fall of 1986, just four months after his appointment to the Supreme Court, Antonin Scalia was faced with a tricky numerical problem. Scalia, a firm believer in capital punishment, was presented with some disturbing evidence that the way the death penalty is administered in the United States—or at least in Georgia—was unconstitutional.

The evidence was at the center of a case that had worked its way up to the Supreme Court, *McCleskey v. Kemp.* The petitioner, Warren McCleskey, had been convicted of shooting a police officer in the face with a .38 pistol, killing him, during the robbery of a furniture store. McCleskey had been sentenced to death.

McCleskey's case had nothing to do with innocence. He was certainly guilty of participating in the armed robbery; he admitted as much, though he denied that he was the triggerman who murdered the policeman. It was more subtle than that. McCleskey argued that he shouldn't get the death penalty because in Georgia, where he was convicted, the application of the death penalty was racist. And he had the numbers to prove it, thanks to University of Iowa professor David Baldus, who analyzed nearly twenty-five hundred murder cases in Georgia looking for patterns.

He found them. Baldus used regression analysis in an attempt to figure out what factors were responsible for a Georgia criminal's getting the death penalty. It was a massive endeavor—looking at such elements as whether the convict had resisted arrest, whether the victim was stabbed multiple times, whether the convict had raped the victim—all sorts of circumstances that might influence a prosecutor to seek the death penalty or a judge and jury to impose it. Like all regression analyses (especially ones with large numbers of variables), it was far from perfect. However, when Baldus crunched the numbers, one factor stuck out as an obvious issue: race.

Prosecutors were much more likely to seek the death penalty if the victim of the crime was white than if the victim was black. Quite naturally, this meant that anyone convicted of killing a white person had a higher probability—about four times as great—of getting the electric chair than someone who killed a black person. It was stunning evidence of racial bias in the courtrooms. There was other evidence too: black convicts were more likely to get the death penalty than white convicts, especially when it came to black-on-white crime. (However, the data supporting this were a little murkier, especially since the statistics for convictions before a key death penalty court decision in 1972 were rather different from those for convictions after the decision.) Though Baldus's study had some flaws, it was extremely disturbing evidence that justice wasn't colorblind.

To an ardent supporter of the death penalty like Scalia, the Baldus study was a serious threat. The moment the Court recognizes that the death penalty is administered in a racially discriminatory manner, it becomes unconstitutional. The Fourteenth Amendment guarantees that each citizen is treated equally by the law, and if African Americans are getting different penalties in court, then it's an obvious violation.

What happened next was a stark example of the proofiness that the Court—particularly Justice Scalia—uses to keep the death penalty alive in the United States. In a five-to-four decision, the Court threw cold water on the Baldus study. Relying on the Georgia district court's technical (and sometimes misguided) cherry-picking of problems with the regression analysis, the majority cast doubt on whether Baldus had found any real evidence of racism. Even if he had, it might not be solid enough proof. The Court decided that "to prevail under the Equal Protection Clause, McCleskey must prove that the decisionmakers in *his* case acted with discriminatory purpose. He offers no evidence specific to his own case that would support an inference that racial considerations played a part in his sentence. Instead, he relies solely on the Baldus study." That is, it's not enough for McCleskey to show that the deck is stacked against him; he has to prove that the dealer cheated specifically in the hand that he'd been dealt. This is questionable jurisprudence, to say the least.

In a moment of candor, the Court admitted that McCleskey's case had to fail, in part, because "McCleskey challenges decisions at the heart of the State's criminal justice system." If the Baldus study was correct, it would have grave implications for Georgia's courts; it would reveal a pervasive racism that undermined many of the courts' decisions over the years. Such a conclusion was so damaging that the justices had to close their minds to the possibility; they had to construct an alternate reality in which the Baldus study was fatally flawed and Georgia's system of justice was completely fair regardless of the race of the victim or the defendant.

Scalia didn't write the opinion. Justice Lewis Powell did the honors. However, Scalia's role is worthy of special attention because of a memorandum that came to light a number of years after the *McCleskey* decision. In the memorandum, dated January 6, 1987, a few months before the decision was handed down, Scalia expressed his opinion that the Baldus study was valid, and that the courts were indeed racist:

> Since it is my view that the unconscious operation of irrational sympathies and antipathies, including racial, upon jury decisions and (hence) prosecutorial [ones], is real, acknowledged by the [cases] of this court and ineradicable, I cannot honestly say that all I need is more proof.

This was a shocking statement. In Scalia's mind, the matter was closed. It didn't matter how much proof you shoveled in his direction; he had already accepted the fact that the Georgia courts' administration of the death penalty was racially biased. But that fact didn't change his mind. The machinery of death must continue to chug away unhindered. But rather than admit that, Scalia joined in with Powell's majority, attacking Baldus's study and requiring higher levels of proof than Baldus's numbers were able to provide. Scalia's memo shows the level of cynicism that the freshman justice was capable of. Even though he believed Baldus's study, he attacked it, (in my opinion) cherry-picking by highlighting its flaws and ignoring its merits. Even though he knew that Georgia's capital punishment system was racist, he demanded impossible levels of proof from a defendant who was fighting for his life. Scalia's memo pulls the mask off of his proofiness.

McCleskey went to the electric chair on September 25, 1991. Powell would step down from the court a few months after penning the *McCleskey* opinion. A few years later, a biographer asked him whether he'd change his vote on any decision. The immediate answer was, "Yes, *McCleskey v. Kemp*." Powell, like some of his fellow justices, had come to the conclusion that the death penalty was not viable in our deeply flawed system of justice. In the years since Baldus's analysis, the case for racial bias has gotten stronger. A number of other studies came out that showed similar patterns to Baldus's findings in death penalty cases. A 1990 government report, which looked at twenty-eight studies, showed in recent times "a pattern of evidence indicating racial disparities in the charging, sentencing, and imposition of the death penalty."

Scalia, for his part, remains unmoved by additional data. If anything, his defense of the death penalty has gotten more strident—and more dependent upon proofiness than ever.

Twenty years after *McCleskey v. Kemp*, Scalia had outlasted all of his peers except for his liberal rival John Paul Stevens. In those years, Scalia used phony mathematical arguments a number of times. In 2008, two researchers caught Scalia in the act. They publicly castigated him for misusing their work in his opinion in *Baze v. Rees*.

In *Baze v. Rees*, convicted murderer Ralph Baze sued to stop his execution on the basis that lethal injection was cruel and unusual punishment. (If so, it would violate the Eighth Amendment.) By a vote of seven to two, the Court rejected the claim. However, it sparked an argument between Scalia and Stevens. Even though the two justices both were in the majority and agreed that lethal injection didn't violate the Eighth Amendment, they fought about whether capital punishment had a deterrent effect—whether the specter of the death penalty keeps some criminals from committing capital crimes.

Stevens was skeptical that capital punishment was an effective deterrent. "The legitimacy of deterrence as an acceptable justification for the death penalty is also questionable, at best," he wrote. "Despite 30 years of empirical research in the area, there remains no reliable statistical evidence that capital punishment in fact deters potential offenders." To Scalia, these were fighting words. Quoting a recent article in the *Stanford Law Review*, he wrote that "Justice Stevens' analysis barely acknowledges the 'significant body of recent evidence that capital punishment may well have a deterrent effect, possibly a quite powerful one.'"

In my opinion, Scalia was cherry-picking data; he seems to have ignored that there was plenty of contrary evidence presented as well. In the very same issue of the journal, the authors of the study acknowledged that they couldn't conclude that capital punishment has in fact been shown to have a deterrent effect. The case for deterrence was at best ambiguous. However, Scalia chose to spin the data to make it look as if the preponderance of the evidence showed a deterrent effect. Shortly after the *Baze v. Rees* decision, one of the article's authors called him out. Cowriting a *Washington Post* op-ed, he accused Scalia of misrepresenting his research: "In short, the best reading of the accumulated data is that they do not establish a deterrent effect of the death penalty."*

It was a rare humiliation for Scalia. He was caught red-handed, picking cherries—by the authors of the study he had raided, no less. But *Baze v. Rees* is not the most egregious example of Scalia's tendency toward proofiness. His opinion in another death penalty case, *Kansas v. Marsh*, is a stunning example of judicial nonsense—a case where a big honking numerical lie was used to send a prisoner to his death.

Michael Lee Marsh was convicted of a particularly brutal crime. He entered the home of Marry Ane Pusch and lay in ambush, waiting for her to return home. When she did, he shot her and slit her throat. Upon leaving, he set the house on fire, leaving Pusch's toddler to be burned alive. He was convicted and sentenced to death.

Marsh's case hinged upon a technicality about sentencing guidelines—how the state of Kansas, where Marsh was convicted, weighed mitigating and aggravating circumstances to the crime. In 2006, the Supreme Court reversed a lower court ruling that had given Marsh a new trial. By a vote of five to four, the justices reinstated a death

* The same op-ed dinged Stevens too, correctly pointing out that the absence of evidence for a deterrent effect is not the same as evidence for the absence of a deterrent effect. In other words, just because you don't have proof that capital punishment works doesn't mean that it's not working.

penalty that had been set aside. In their dissent, the four outvoted justices cited some evidence that improper convictions of felons were "remarkable in number, and they are probably disproportionately high in capital cases," including a recent article in the *Journal of Criminal Law and Criminology* that listed 340 prisoners (not all on death row) who had been exonerated between 1989 and 2003.

Scalia launched a violent counterattack on what he termed the sanctimonious finger-wagging of the anti-death-penalty forces. The foundation of his argument was an outrageous bit of proofiness. Scalia asserted that the justice system was almost completely foolproof: that its error rate for felonies was a mere 0.027 percent. It's a mind-blowing figure—if true, it would mean that juries and judges have a nearly superhuman ability to prevent the innocent from winding up behind bars.

The figure came from an op-ed piece that had appeared in the *New York Times* six months earlier. Penned by Joshua Marquis, a prosecutor from Oregon, it attacked the 340-exonerations study, and generated the 0.027 percent figure by a back-of-the-envelope calculation:

> So, let's give the professor the benefit of the doubt: let's assume that he understated the number of innocents by roughly a factor of 10, that instead of 340 there were 4,000 people in prison who weren't involved in the crime in any way. During that same 15 years, there were more than 15 million felony convictions across the country. That would make the error rate .027 percent—or, to put it another way, a success rate of 99.973 percent.
>
> Most industries would like to claim such a record of efficiency.

This is a stunningly incorrect calculation. The main problem is that it's fallacious to look at all 15 million felony convictions across the country in coming up with an error rate. It's comparing apples to oranges, because most of those felonies never had any chance of being overturned, even if they were wrongful. To calculate the real error rate, you have to look very carefully at which of those 15 million felony convictions you should include in the calculation. It's not an easy task.

For one thing, not all felonies are created equal. People who are convicted of check-kiting, blackmail, and other nonviolent felonies tend to get out of jail relatively quickly; it's the violent offenses like rape and murder that get the long sentences. Convicts with long sentences are the ones who are going to get aid from pro bono lawyers and from groups like the Innocence Project, and perhaps get their convictions overturned. Rapes and murders are just a tiny percentage of the felonies committed annually (in 2002, 0.6 percent of felony convictions were for murder and 3.3 percent were for rape or sexual assault). However, murder, rape, and sexual assault accounted for 326 out of the 340 exonerations listed by the *Journal of Criminal Law and Criminology* paper. That is, the vast majority of exonerations were drawn from less than 4 percent of the pool of felonies. These 600,000 people convicted of murders, rapes, and sexual assaults are the ones whose cases are really eligible for being overturned—by virtue of a high level of scrutiny and long sentences, these are the types of felony convictions that should go into Marquis's calculation, not the 15 million felony convictions of all kinds.

However, even this 600,000 felonies figure is deceptive. It includes lots of people who never went to trial in the first place. If you want to find out the proportion of people who are wrongly convicted, you should only look at those convicts who claim that they're innocent, rather than the ones who admit their guilt. In 2002, more than 90 percent of people who went to jail for rape and sexual assault entered a guilty plea; nearly 70 percent of murderers did the same thing and admitted their guilt. An error rate should exclude these felons, leaving only about 80,000 felons eligible to be exonerated.

This is obviously quite a bit smaller than the 15 million that Marquis originally used. Repeating the calculations, substituting the 80,000 number for the wildly incorrect 15 million one, yields an error rate of around 5 percent, not 0.027 percent. Perhaps one in twenty convictions is wrongful.*

This is just a ballpark figure, of course, but it's probably not so far from the truth. A handful of studies have attempted to come up with a firm number for wrongful convictions, and they yield similar figures. For example, a 2007 paper looked at convicts who had been convicted of raping and murdering a victim and then were subsequently cleared by DNA evidence—a sure sign of a wrongful conviction. The data implied that between 3 and 5 percent of those convictions were in error.

Scalia's image of a nearly perfect judiciary, one that falsely convicts a felon only 0.027 percent of the time, is absurd. He knows it too. Scalia is sophisticated enough to understand how flawed the judicial system must be. He no doubt realizes that a human endeavor

---

* This is something of an oversimplification, because at least one person who has been exonerated through DNA evidence did in fact plead guilty to try to get a lesser sentence. This means that, strictly speaking, we shouldn't exclude *all* of the admissions of guilt; we should add back the ones who admitted guilt but knew they were innocent. This would reduce the number to a bit less than one in twenty. (These numbers are obviously very sloppy—don't forget the factor of ten thrown in as a rhetorical device—and shouldn't be taken as anything more serious than a really rough guess.)

as messy and subjective as a criminal trial would not be correct 99 percent, much less 99.97 percent, of the time as he claimed. Scalia was almost certainly aware that Marquis's figure was a Potemkin number, but he used it anyhow.

Scalia's proofiness served a definite and cynical purpose. The ugly fact that the justice system is imperfect, the possibility that as many as 5 percent of the people who are on death row are not guilty of the crime they committed, is damning for proponents of capital punishment. The sheer number of errors in our courts make the finality of an execution—of an irreversible mistake—hard to justify. The only way around that hard fact is to pretend it isn't so. Scalia used his Potemkin number to construct an alternate reality. In this parallel universe, the justice system is nearly flawless. No undeserving innocent is ever put to death. And each night, judges sleep with dreams untroubled.

8

Propaganda by the Numbers

Modern man needs a relation to facts, a self-justification to convince himself that by acting in a certain way, he is obeying reason and proved experience. . . . The problem is to create an irrational response on the basis of rational and factual elements. That response must be fed with facts, those frenzies must be provoked by rigorously logical proofs.

—Jacques Ellul, *Propaganda*

Every afternoon at five o'clock, in a reinforced building a few blocks from the Saigon River, the Pentagon pelted the press with numbers, hoping to convince them that the United States was winning the Vietnam War.

It wasn't enough for generals to declare that the war was being won. Reporters were far too cynical about human nature to take a military official at his word. Journalists demand objective facts, verifiable truths that they can use to prop up a sagging story. So the Pentagon began providing them.

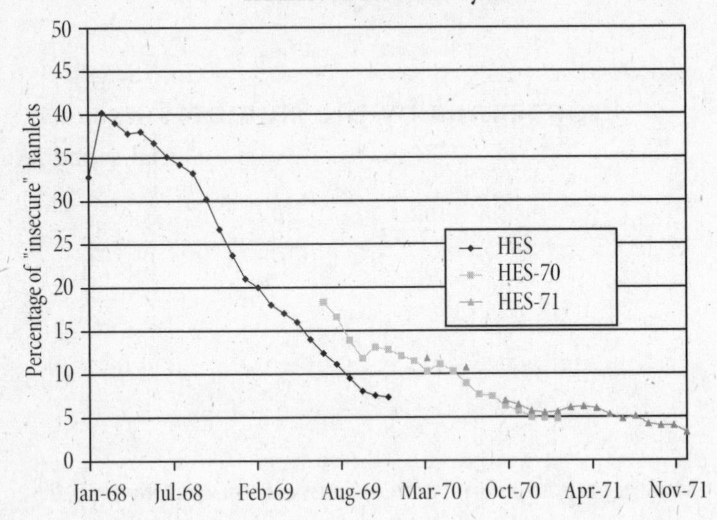

Figure 16. The United States marches toward victory in Vietnam, according to the Hamlet Evaluation System.

The briefings were a jumble of numbers and statistics. Officers and officials recited a litany of body counts, numbers of weapons captured, head counts of troops in different theaters—any quantifiable information that could help prove that American forces were making progress, no matter how ridiculous the numbers were. The Pentagon invented a "Hamlet Evaluation System" to quantify how much of the South Vietnamese countryside was "pacified"—under U.S. and South Vietnamese control—and proudly reported the numbers to the press. Each month, the numbers ticked up, showing that government forces were slowly winning the war. Year after year, the number of hamlets under control dutifully kept climbing and the number of Vietcong-controlled hamlets kept dropping, demonstrating slow and steady progress until our inevitable defeat burst the balloon.

The savvier journalists, the ones who got away from Saigon and visited the troops in the field, saw that the situation was steadily deteriorating, not getting better. They realized that official-sounding statistics dished out at the daily briefings, which they quickly dubbed the "five o'clock follies," were fantasy. As *Time* magazine put it in 1969:

> Time and again, military briefers in Viet Nam have "proved" that the war was being won with the help of impressive "body counts" of enemy dead that were impossible to verify, let alone dispute. With the aid of computers, U.S. officials were equally sanguine about stating to the decimal point how many villages were "secure" in Viet Nam.

No matter how indisputable the mess-up, no matter how obvious it is that somebody has screwed the pooch, a statistic can put a happy face on failure. Say you're a general who's just tested a very expensive missile and it has failed to hit its target for the sixth time in a row. Do you tell the press that the test was a failure? Of course not! Simply say that the test was a success because the missile had successfully achieved sixteen out of seventeen objectives. (Of course, the seventeenth objective was actually hitting a target.) That way, you can say with a straight face that you were "encouraged" by the test.*

---

* Yes, this actually happened. The missile in question was an anti-ballistic-missile rocket known as the THAAD, the test occurred on March 29, 1999, and the general was Lieutenant General Paul Kern of the U.S. Army's Acquisition Corps.

At times, the Pentagon's attempts to use bogus statistics—Potemkin numbers and other forms of proofiness—to spin unpleasant facts might seem almost comically inept. But they're a surprisingly effective tool for controlling the press. Proofiness uses a reporter's greatest strength, his reverence for objectivity and truth, against him.

"If your mother says she loves you, check it out."

It's an old journalistic saying, often attributed to the Chicago City News Bureau, repeated over the years by editors and journalism professors all around the country. It encapsulates a journalistic ideal: a mistrust of anything that isn't observable or verifiable, a rejection of hearsay. Responsible reporters must take nothing for granted—only by basing every single sentence of your story upon observations or verifiable facts can you be assured that you're reporting the truth. Everything that isn't bolstered by an observation or another sort of fact should be set aside. As a Columbia Journalism School professor puts it:

> I was trained to believe in only what is observable and quantifiable. . . . Journalists feel most secure with the batting average, the stock price, the body count, the vote tally (well, maybe not in Florida in 2000). We feel comfortable watching the Rose Bowl game or the ticker-tape parade unfold before our eyes.

Journalists are supposed to retreat into a fortress where we admit nothing but the information that we get from our own senses and the verifiable truths that statistics seem to represent. This sort of information is the raw fuel of journalism. It is what gives journalists their voices; without hard facts to pin our words to, we are powerless to express ourselves.

The Pentagon realized this long ago. During Vietnam, it would toss out Potemkin numbers and fruit-packed data week after week, providing raw material for journalists who needed to produce stories for their editors. Though many accomplished journalists scorned the five o'clock follies and got their information by talking to troops in the battlefields, others were content to stay in Saigon and regurgitate the phony statistics that the Pentagon spoon-fed to them. What's more, these statistics didn't even need to be terribly believable for them to serve their purpose. The reason is counterintuitive: the lies are not meant for ideological enemies, but for allies.

Weekly body counts and monthly hamlet evaluation numbers were never likely to convince the war skeptics in the journalistic community—they were intended to be used by hawkish journalists and pundits, the ones who were already convinced that the war was being won, such as Joseph Alsop. Alsop's "Matter of Fact" column would occasionally showcase the Pentagon-created numbers that proved that the war was being won—harvest statistics, body counts, numbers of captured documents. Alsop deployed even the Hamlet Evaluation System to paint a rosy picture of the war:

> One would like to cite countless facts about this process of [pacification], which has brought the number of people under government control in the delta from less than 3 million to more than 4.6 million in less than a year; and has meanwhile reduced the [Vietcong]-controlled population from nearly 2 million to just above half a million with another half-million "contested."

This was in late 1969; at the time, even the Nixon administration was not as sanguine about the prospects of success as Alsop was.*

The Pentagon is only one of a vast array of proofiness factories, which manufacture phony statistics and fake facts and cast them into the air in hopes that some ally will make use of them. The strategy often works, even with the most ridiculous Potemkin numbers imaginable. No number is too silly.

In March 2007, a marketing research group created a report for its auto industry clients that came to a very surprising conclusion. Driving the notorious gas-guzzling Hummer H3, they claimed, was better for the environment than driving the energy-efficient Toyota Prius hybrid.

Even a brief look at the report showed that it was chock-full of Potemkin numbers. For example, it claimed, without much justification, that the average Prius would be driven 9,146 miles a year for 11.92 years, clocking a grand total of 109,000 miles. The Hummer H3, on the other hand, would be driven much harder and last much longer before it gave up: 13,543 miles a year for 15.28 years, 207,000 miles in all. (This was nothing compared to the Hummer H1, which would be on the road for an incredible 34.96 years and 379,000 miles *on average*.) These are clear Potemkin numbers—there's no way that the typical Hummer H1 would last for nearly 400,000 miles—and they have the effect of inflating the energy efficiency of the Hummers and deflating the energy efficiency of the Prius.

The report was transparent nonsense. But it was music to the

* As one reporter put it in 1969, "In the political backrooms, they talk about getting an 'Alsop-proof' peace formula. That means a formula which columnist Joseph Alsop would accept, albeit grudgingly."

ears of anti-green pundits and global warming skeptics. Rush Limbaugh trumpeted the report on his radio show, and it began to gather a buzz. Within a few weeks, it hit the mainstream. In April, *Washington Post* columnist and inveterate global warming denier George Will crowed about the study: "Speaking of Hummers, perhaps it is environmentally responsible to buy one and squash a Prius with it," he wrote, repeating the Potemkin numbers. For the next half year, every time a pundit wanted to take a swipe at the green movement, all he needed to do was trot out the report and build a column around it.

Even though the report's numbers were bogus, they served their purpose. The marketing group helped its clients by giving ammunition to allies of the auto industry. Antienvironmentalist pundits cast doubt upon the wisdom of buying energy-efficient cars. Arguing about whether the Hummer is greener than the Prius is a phony debate—one side is based upon a highly dubious premise—yet the argument was aired again and again in the press.

This is in part because phony debates are another journalistic Achilles' heel. Reporters are trained to present a neutral point of view, presenting both sides of an argument in a balanced manner. However, when the argument is lopsided, with the vast preponderance of the evidence supporting one side over the other, the reporter's "balance" is just as phony as the debate—the journalist tends to give too much credence to proofiness-bolstered fringe ideas at the expense of those in the mainstream. No matter how ridiculous one side of the argument is, no matter how dependent it might be upon proofiness, the press dutifully broadcasts it and amplifies it, giving manufactured "facts" a life of their own.

The press is just one target of proofiness. Manufacturers of Potemkin numbers, cherry-picked studies, and other forms of numerical nonsense have plenty of other people interested in their wares. In the right hands, a choice piece of proofiness can be a powerful tool to gain ground in even the most hard-fought political battles.

In the United States, for example, the debate over abortion has been bitter for decades. The 1973 Supreme Court decision in *Roe v. Wade* wiped out state laws that barred abortions; since then, a number of states have been attempting to regain lost ground by putting up barriers that make abortions harder to get.

In Texas, the state where "Jane Roe" and Henry Wade came from, abortion providers are required to give every potential patient a piece of proofiness in hopes of getting her to change her mind and keep the baby. A 2003 law—the "Woman's Right to Know Act"—dictates that every abortion doctor tell each client about "the possibility of increased risk of breast cancer following an induced abortion." The problem is that there is no increased risk. The association between abortion and breast cancer is nonsense.

Over the years, a handful of studies seemed to show a slight increase in breast cancer rates among women who have had abortions, but the subject really caught fire in 1996, when endocrinologist Joel Brind analyzed two dozen studies that had data about cancer as well as abortions. After crunching the numbers, Brind concluded that a woman who had an abortion had a 30 percent greater chance of contracting breast cancer than other women. However, Brind failed to take into account a number of biases that create the appearance of a pattern when there really isn't one. For example, women who are perfectly healthy are much less likely to report a prior abortion than women who are gravely ill—the ones who are sick are much more likely to be frank about their medical history.

(In one Dutch study, researchers found that an even more dramatic result than Brind's—a 90 percent increase in breast cancer risk among women who'd had abortions—essentially disappeared when you looked only at people in more liberal parts of the country and who were thus more likely to admit to a prior abortion.) A number of larger, more carefully analyzed studies that have come out since Brind's analysis have shown no increased risk of breast cancer from having an abortion.

The link between abortion and breast cancer is randumbness; people saw a pattern when there really wasn't one. In 2003, the National Cancer Institute held a conference on the subject—at the end of the meeting, the report stated that it was well established, scientifically, that abortions did not increase the risk of breast cancer. (The report was endorsed by all of the conference participants except for one: Joel Brind.) The scientific consensus was as clear as it could be; the link was fictional. Yet a few months later, Texas passed the Woman's Right to Know Act. Abortion doctors around the state were forced to put a phony doubt into their patients' minds—doubt about whether they'll get breast cancer because of an abortion.

Texas isn't alone. Under the guise of protecting patients, abortion foes have enacted laws in five other states that make abortion doctors tell their patients about the possible link between abortion and breast cancer. Even though the best current scientific knowledge says that the link doesn't exist, it serves a political purpose to make that risk loom large. It's risk mismanagement codified into law, intended to scare women into changing their minds about getting an abortion. And that's not the only form of proofiness that legislators throw at potential abortion recipients via ironically named "right to know" laws. South Dakota's abortion law is a good example; it dictates that a doctor must tell the patient about an "increased risk of suicidal ideation and suicide." Here too, mathematical nonsense has been turned into the law of the land.

It *is* true that women who have abortions have a higher risk of suicide than those who don't. According to a 2005 Finnish study, women who terminate their pregnancies are about six times more likely to kill themselves. However, the same study also found that women who have abortions have a higher risk of being murdered—ten times more likely than those who went through with the birth. They were also 4.4 times more likely to die in an accident. These statistics hint at what's really going on here: causuistry. It's not that abortions are causing people to commit suicide, any more than abortions are causing people to get murdered or have accidents. Women who are in a high-risk category for suicide, homicide, and accidents are almost certainly more likely to get into a situation where they need an abortion than those in the general population. Mentally ill people, risk-takers, women who run with a wild crowd—they are more likely to get into trouble of all sorts, including having an unwanted pregnancy. Conversely, women who have a child are more likely to be more stable and responsible, and less likely to take needless risks. Yet antiabortion lawmakers turn that correlation into causation, forcing doctors to scare their patients with the idea that going through with the abortion might cause them to commit suicide later on. The "right to know" laws have nothing to do with knowledge. They are vehicles for whatever bits of proofiness can be deployed to support the lawmakers' opinions.

The left is guilty too. Liberals get every bit as lathered up about guns as conservatives do about abortions, so they manufacture proofiness to try to get their way. In 1996, a young historian at Emory University, Michael Bellesiles, burst onto the academic scene.

His claim to fame was an article in the *Journal of American History* that argued that guns were rare in eighteenth- and nineteenth-century America, and that the populace resisted the government's attempts to arm its citizens. If true, it would change the tenor of the debate about gun control and the right to bear arms; it would reframe the idea of a gun-toting public, making it a relatively modern invention rather than an organic part of American culture. The paper won Bellesiles a great deal of praise and attention—as well as a book contract. Bellesiles's magnum opus, *Arming America*, was lauded by the *Economist* and even won the prestigious Bancroft Prize. It was quickly becoming a very influential piece of research, and was embraced by antigun activists, even though the research turned out to be mathematical tripe.

Bellesiles claimed to base his conclusions upon numbers of guns found in archival documents—particularly wills and other probate records from around the country. But it soon became apparent that the data were questionable at best. He had clearly misinterpreted some data and cherry-picked others. There were also dark intimations that Bellesiles was making up some numbers out of whole cloth. One critic pointed out that one set of Bellesiles's numbers was mathematically impossible, implying that Bellesiles was artificially deflating the number of guns in the region he was studying. When Emory University investigated Bellesiles's work, they declared that the professor was guilty of either "extremely sloppy documentation" or fraud—and seemed to lean toward the latter, finding "evidence of falsification" of some of his data. In other words, Bellesiles had probably committed the ultimate academic sin.* His desire to

* An allegation that Bellesiles adamantly denied.

change the debate on gun control had led him to manufacture Potemkin numbers. His book was withdrawn by its original publishers and he was forced to resign.

Both the left and the right have embraced proofiness, as it has proven its power. No matter how moronic a number might be, no matter how transparently false a statistic is, it can still justify a piece of legislation or a public policy . . . or a false belief.

Proofiness gives fantasy the appearance of fact, reinforcing it, strengthening it so that it can resist assault from without. In some sense, proofiness is the antithesis of knowledge.

In the first half of the twentieth century, scientists at Bell Labs and elsewhere created a new branch of science known as information theory. Originally conceived as a way to answer some very mundane telecommunications problems (such as how many telephone calls can be crammed onto a single copper wire), information theory soon grew far beyond its roots and became one of the most profound scientific theories of the century. It explained how information gets transmitted, stored, and manipulated by information processing devices: devices like computers—or our brains.*

One of the dicta of information theory is that information resides in the unexpected. We gain knowledge when we encounter what we don't anticipate. A stream of data that we can predict with perfect accuracy contains no information; it can't tell us anything that we don't already know. The quest for knowledge is a quest for novelty, a search for a new set of data or a new idea that forces us to

* For those craving more detail about the interesting (and disturbing) implications of information theory, my book *Decoding the Universe* might appeal.

look at the world in a slightly different way than we did before. Knowledge-gathering is systematic demolition and reconstruction of our view of the world.

It can be an unsettling and uncomfortable process. It's never easy to destroy a cherished myth, to abandon a deeply held belief, or to inject shades of gray into a debate that once seemed black and white. It's human nature to resist change, to cling to our old, familiar ideas instead of abandoning them in the face of new information. We shrink away from data that challenge our prejudices; we tend to seek out—and to believe—data that reinforce them.

This is nowhere more apparent than in the news media. A few decades ago, we had only a small handful of sources from which we could get our news. Over breakfast, we would read a newspaper or two. On our drive to and from work, we could choose to listen to two or three news radio stations. After dinner, we could watch the television news on three or four channels. In bed, we might relax with a magazine. With the advent of cable television and then the Internet, the number of news outlets proliferated enormously. The audience fragmented and then atomized. Gone were the days when Americans were forced to pick one of three nightly news broadcasts. Now there are so many outlets that we suddenly have the ability to find the source of news that makes us the least uncomfortable.

More and more, people seem to be seeking out news outlets that reinforce their beliefs without challenging them. Conservatives can now get their facts from Fox News; liberals can go to the Huffington Post for theirs. We no longer have to confront ideas that force us to reevaluate our positions. Instead, we can only listen to the ones we already agree with. We can wallow in our myths, undisturbed by the inconvenience of doubt. Gaining knowledge need no longer be the uncomfortable by-product of listening to the news.

Proofiness is at the center of this problem. Phony numbers have the appearance of absolute truth, of pure objective fact, so we can use them as a justification to cling to our prejudices. Fruit-packing can give an aura of respectability to the most wrongheaded notion, while a cleverly deployed Potemkin number or a disestimate can stave off even overwhelming evidence that an idea is incorrect. Proofiness is the raw material that arms partisans to fight off the assault of knowledge, to clothe irrationality in the garb of the rational and the scientific. This is what makes it such a powerful tool for propaganda.

In the United States, propaganda is a way of undermining democracy. It lets demagogues whip up a storm of irrational emotion, a thoughtless frenzy that leads people to vote against their interests and to support policies that they would otherwise reject. It is a subtle form of mind control, a mechanism for tricking people into agreeing with their leaders. Just as proofiness undermines democracy in other ways—diluting our votes, disenfranchising our citizens, prejudicing our justice system—through propaganda, it can rob us of our democratic right to think for ourselves.

Not all propaganda need make use of proofiness. Indeed, arguably the most effective modern use of propaganda didn't have much of a proofiness component to it. In the months leading up to the Iraq war, neoconservative government officials such as Richard Perle enlisted the aid of a *New York Times* reporter, Judith Miller, to spread stories about fictional Iraqi projects to build weapons of mass destruction. The administration used nonexistent weapons as the justification for invading Iraq, and by the time the lies were uncovered it was far too late.

However, masters of propaganda know that proofiness is arguably the best weapon in their arsenal. It's subtle. If the proofiness is manufactured by an expert, it can be very hard to discover. A well-crafted number deployed in precisely the right manner can move American—and world—opinion in a favorable direction. And it's extremely hard to combat, especially if there aren't enough data to prove that the number is fake. A smart proofiness-based propaganda campaign can remain effective for decades. For example, in the 1970s and 1980s, a group of hard-liners in the Pentagon was responsible for an extremely crafty propaganda offensive based upon a choice piece of proofiness, one that all but obliterated a Democratic push for a nuclear test ban treaty with the Soviet Union.

On July 20, 1982, the front page of the *New York Times* contained the surprising story that the United States was abandoning nuclear treaty talks with Russia. It was a carefully controlled leak; government officials had contacted a *Times* reporter to break the news to the world and to explain the rationale for what might seem like a bellicose act. The next day, a follow-up story in the *Times* explained the administration's position: the Soviet Union was probably cheating on a prior nuclear treaty, so there was no point to signing a new agreement. The accusation was quite specific; an unnamed government official was accusing the Soviet Union of likely violating the Threshold Test Ban Treaty, which dictated that nuclear tests could be no larger than 150 kilotons—about ten times the size of the bomb that leveled Hiroshima. "On several occasions seismic signals from the Soviet Union have been of sufficient magnitude to call into question Soviet compliance with the threshold of 150 kilotons," the (unnamed) official told the *Times* reporter. Within a few days, the rhetoric got a little stronger. The *Times* repeated one (again unnamed) government official's assertion that "several" Soviet tests "had been estimated at 300 kilotons." These accusations were specific and credible, and they were devastating to Democratic hopes of signing a comprehensive test ban treaty with the Soviet Union. They were also utterly false. The accusations were a tricky bit of proofiness in an obscure equation used to estimate the size of nuclear blasts.

It's a really tough task to figure out the size of an underground blast halfway around the world. The best method is to listen with seismographs—the same instruments that detect earthquakes—and from the violence of the shaking, you can get a rough estimate of how big an underground blast might be. Over the years, geologists got pretty good at making these estimates. After observing many experiments, they came up with an excellent equation for calculating the yield of a nuclear weapon detonated at the Nevada Test Site based upon how violently the earth shook:

$$m_b = 3.92 + 0.81 \times \log(Y)$$

where m_b is the Richter scale of the earthquake caused by the test and Y is the yield of the nuclear explosive in kilotons. This was how the government used to calculate the yield of nuclear tests in the Soviet Union. Based upon the violence of the earthquakes, the Reagan administration concluded that the Russians were detonating 300-kiloton bombs, in direct violation of the treaty.

The only problem was that the equation was wrong. The precise shape of the equation depends on a lot of factors, particularly the geology of the site where the blast occurs. A blast in hard bedrock will shake the earth much more than the same-sized explosion in loose, silty soil. Similarly, seismographs will have an easier time picking up an explosion in a geologically stable region than in an active region that's crisscrossed by faults. Nevada is the latter; it is in a relatively active region of the earth with loose tuff and alluvial soils, while the Semipalatinsk test site in the Soviet Union is in hard basaltic rock in a relatively stable region of the planet. As a result, the Russian equation had a similar form, but it was calibrated somewhat differently from the Nevada equation:

$$m_b = 4.45 + 0.75 \times \log(Y)$$

This means that a 6.0 earthquake in Russia would be caused by a bomb of about 100 kilotons, rather than the nearly 400-kiloton whopper that would produce the same-sized temblor in Nevada.

The alleged Soviet violations of the Threshold Test Ban Treaty were a result of deliberately using the wrong equation to calculate the yield. In the late Nixon administration, a group of Pentagon hard-liners opposed to arms control deals with the Soviet Union began to manufacture evidence that the Russians were violating the treaty. "People started circulating calculations of yield based upon the Nevada curve," which artificially inflated the apparent size of Soviet explosions, says Fred Lamb, a physicist at the University of Illinois at Urbana-Champaign. "Of course, some explosions seemed to be in violation of the treaty, and this caused people to go crazy." Even though scientists complained that the numbers were wrong and that the "violations" were nonsense, they were ignored. "It became clear to me that [the false numbers] were being used by people opposed to arms control agreements in general," Lamb says.

The matter came to a head in the Reagan administration. Officials within the Pentagon, particularly the hawkish assistant secretary for international security policy, pressed President Reagan to make louder and louder complaints about the Soviet Union cheating on nuclear treaties.* Behind the scenes, a handful of seismologists and physicists were trying to get the administration to use the correct equation and to stop making false claims, but the Pentagon quashed their reports, and the assistant secretary launched a vicious attack on the scientists who were trying to get the truth to come out. "[He] actually accused seismologists of being communists and Soviet sympathizers," says Lamb. And, in the meantime, the false allegations against the Soviet Union continued unhindered.

In 1985, Reagan went to Congress with a formal accusation: "Soviet nuclear testing activities for a number of tests constitute a likely violation of legal obligations under the Threshold Test Ban Treaty of 1974." And he harped on it year after year, never relenting on his false accusations. "As evidence continued to accumulate, as the arguments were knocked away one by one, the administration still made a case for a 'likely violation,'" says Columbia University geophysicist Paul Richards. (A joint experiment in 1988 where Americans were allowed to measure a Russian test yield directly proved beyond any doubt that the seismologists were correct all along. The second equation was correct, and the Soviets hadn't violated the treaty.)

Yet the lie was effective propaganda. It gave hard-liners a justification for rejecting peaceful overtures from the Soviet Union, and the dream of a comprehensive test ban treaty—a moratorium on all nuclear testing—was delayed for more than a decade. It only came to be in 1996, during the Clinton administration. Though the president signed the treaty, Democrats had since lost control of Con-

---

* Ironically, the Soviets *were* cheating on their biological weapons treaty—just not on the nuclear test ban.

gress, and the Republican Senate refused to ratify it. As a result, the Comprehensive Test Ban Treaty remains in legal limbo unto this day.

The accusations were a diabolical piece of propaganda. A subtle bit of proofiness allowed America to bear false witness against our Soviet adversaries. It's a classic pattern; indeed, the whole episode looks quite similar to the propaganda that led to the Iraq war. Hawkish government officials, very much like the assistant secretary for international security policy, leaked phony information to a *New York Times* reporter, who then used the newspaper as a platform for self-serving propaganda. It's chilling to realize just how similar the episode was.

The hawkish assistant secretary for international security policy was a neoconservative gentleman named Richard Perle. And the *Times* reporter? Judith Miller.

Proofiness is toxic to a democracy, because numbers have a hold on us. They are powerful—almost mystical. Because we think that numbers represent truth, it's hard for us to imagine that a number can be made to lie. Even the oafish Joe McCarthy knew this; when he declared that 205 communists had infiltrated the State Department, his outrageous falsehood was given the appearance of absolute fact. But proofiness is not merely a tool for propaganda as it was for McCarthy—it is much more dangerous than that. Democracy is a system of government based upon numbers, and rotten numbers are eroding the entire edifice from within. Proofiness in America is used to disenfranchise voters and to bias elections. It's used to drain money from our treasury and put it in the pockets of unscrupulous businessmen. Proofiness tilts the scales of justice and helps condemn the innocent. It puts lies in the mouths of reporters. There's no institution in a democracy that's immune.

The only antidote to proofiness is, ironically, mathematics. Numbers can shatter myths and can disprove falsehoods. They can be turned against their abusers. They can banish proofiness—or at least reduce its influence.

The understanding that real-world numbers come from imperfect measurements can inoculate you against Potemkin numbers, disestimation, and fruit-packing—it imparts a skepticism about where numbers come from, whether they're trustworthy, and whether they've been presented in an honest and straightforward manner. A little mathematical sophistication—and a little practice—allows you to recognize errors of randumbness, causuistry, and regression to the moon; once you get used to spotting phony patterns and false connections, you'll begin to see them everywhere. You'll see how advertisers pump up their products, how bureaucrats cover their failing projects, and how would-be prophets convince the unwary to believe meaningless predictions. And while mathematical knowledge won't stop businesses from ruining the economy, politicians from stealing elections, and court officers from undermining our justice system, it will prevent the malefactors from getting away unobserved.

Mathematical sophistication is the only antidote to proofiness, and our degree of knowledge will determine whether we succumb to proofiness or fight against it. It's more than mere rhetoric; our democracy may well rise or fall by the numbers.

Acknowledgments

My most heartfelt gratitude goes to my editor, Wendy Wolf, my copyeditors, Don Homolka and Roland Ottewell, and my agents, Katinka Matson and John Brockman—without their careful shepherding from idea to manuscript to final proofs, I would never have been able to complete this book.

Thanks to my colleagues at NYU, who have been wonderful to me. Particular thanks go to Pete Hamill for discussions about Vietnam, but also to the other members of the journalism faculty, as well as to the deans who give us the freedom and the support to pursue our interests.

Most important, I am thankful for the love of my family: my parents, my brother, and my wife Meridith—and, last, but not least, my new daughter, Eliza Rose, who gave an author the best incentive ever to meet a deadline.

Appendix A: Statistical Error

Imagine that you're a pollster. It's October 2000—a very exciting time to be conducting polls, because the race between Bush and Gore is neck and neck. You don't know this (after all, you haven't conducted your poll yet), but the country happens to be *exactly* evenly divided; exactly 50 percent want Bush to win and 50 percent want Gore. The truth—the genuine preference of the electorate—is split right down the middle. An ideal poll, one without any errors whatsoever, would show that the two candidates are exactly tied at 50 percent.

Let's conduct a poll and see whether your answer corresponds to the truth. Now, it's impractical to go back in time to 2000 and ask people their preferences. Luckily, there's an excellent substitute for a 2000 voter: a coin. Heads, Bush; tails, Gore.

"Excuse me, sir . . . will you be voting for Bush or for Gore?" Flip the coin—it will come up for Bush 50 percent of the time and Gore 50 percent of the time, just like an exactly split electorate would. In this particular scenario, flipping a coin is a great simulation of asking a person whether he prefers Bush or Gore.

"Heads? Bush, then. Thank you very much for your time."

It costs money to ask people questions, and your company is very cheap. They've only given you the cash to poll sixteen people in all. Flip a coin sixteen times and count up the number of heads and the number of tails.

Did your poll get the right answer? Did you show that the electorate was split right down the middle, with eight people voting for Bush and eight people voting for Gore? Probably not. There's only about a one in five chance that your sixteen answers split exactly evenly, giving you the correct answer. Most of the time, your poll is slightly wrong. Often, you will get nine Bush votes to seven Gore votes, making it look as if Bush has a 56 percent to 44 percent lead over Gore. Similarly, it's quite likely that you get seven Bush votes to nine Gore votes, making it appear that Gore has the equivalent lead. It's also common for a pollster to get ten Bush votes to six Gore votes or vice versa—making it seem as if one of the candidates has a huge 62 percent to 38 percent lead.

If you did the exact same poll over and over (try it!), you'd see that the results jump around quite a bit. Sometimes your poll is dead accurate—it gives exactly the right answer, showing that the candidates are neck and neck. However, more often, through random chance, you happen to get an extra Bush or Gore supporter or two in your sample, artificially making it look as if one candidate has a lead over the other. Fully 80 percent of the time, you're going to get the wrong answer—just because of dumb luck. This is statistical error, and it's an unavoidable consequence of sampling. The laws of randomness inject some uncertainty into the polling process. You're never certain that your answer represents the truth.

Luckily, statistical error is extremely predictable. If you did your

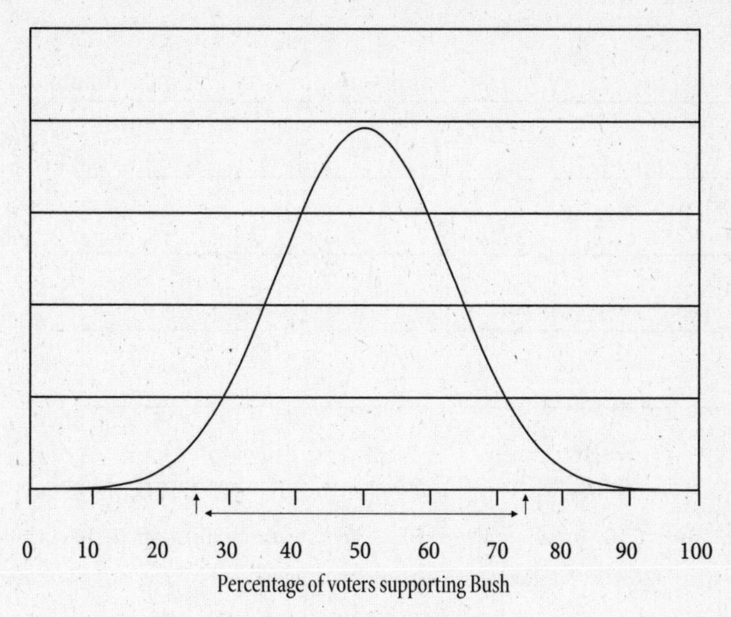

Figure A1. Expected outcomes of polls of 16 people.

poll of sixteen people over and over again and plotted the results, you would see that they followed a pattern—a bell curve (above).

If there were no such thing as statistical error, if your poll result always was a perfect reflection of the truth, it wouldn't be much of a curve at all; it would just be a pencil-thin line sitting on the 50 percent mark. Every single time, your poll would give an answer of 50 percent for Bush, 50 percent for Gore. However, that's not the way the world works. Statistical error ensures that you've got a nice fat curve. Once in a while, your poll gives you the right answer, but most of the time, your results are off—often by quite a bit. The fatter the curve, the more likely it is that your poll is really far away from the truth.

There's nothing you can do to eliminate statistical error, but you can make it smaller. If you can convince your corporate bosses to

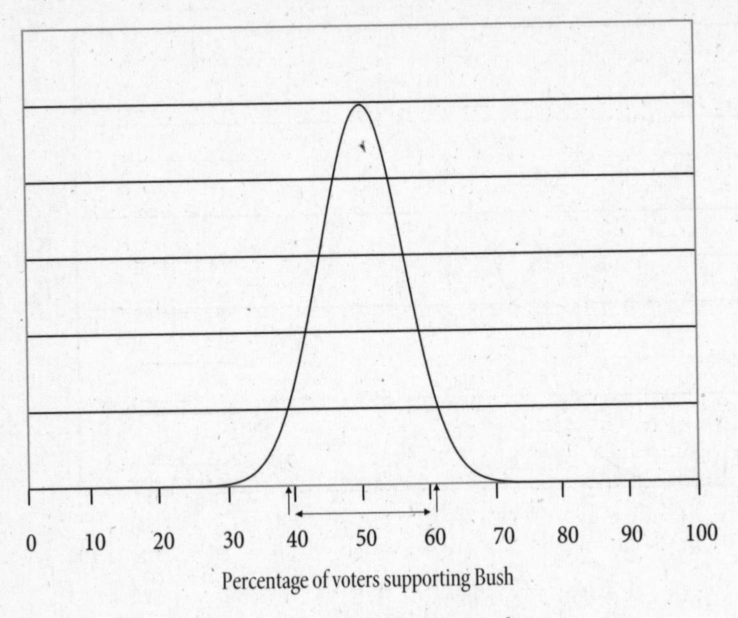

Figure A2. Expected outcomes of polls of 64 people.

shell out a little more money and interview sixty-four people instead of sixteen, the curve gets somewhat thinner (above).

This means that your poll is less likely to be wildly wrong. The more people you poll, the less frequently random errors will throw off your answer by a large amount. Those errors never completely disappear, but you can shrink them to a reasonable size by including lots and lots of people in your poll. If you managed to poll 1,024 people, for example, the curve would look like the one on page 249.

As you can tell, most of the time your poll results will be relatively close to the truth. Unlike your poll of sixteen people, where it was quite common to see results that showed Bush or Gore with a twenty-point lead, that almost never happens with a poll of 1,024 people.

What about margin of error? It's essentially a measure of the

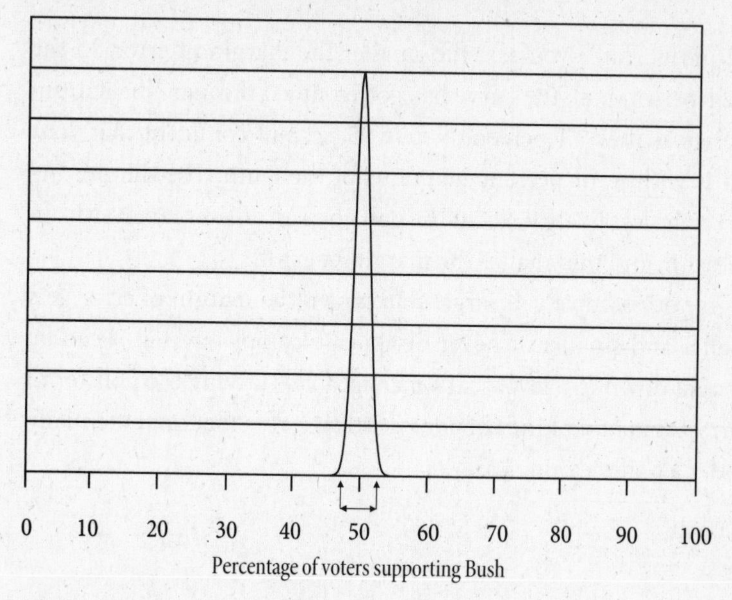

Figure A3. Expected outcomes of polls of 1,024 people.

fatness of the curve. The margins of error are the lines, marked on the preceding graphs, such that 95 percent of the bell curve falls between them. In the sixteen-person poll, roughly 95 percent of the time the answer you get will be somewhere between 74 percent and 26 percent for Bush. This means that there's a 95 percent chance that your answer will be within twenty-four percentage points of the truth that exactly half the population prefers Bush. This is your margin of error: plus or minus 24 percent. Only rarely will your poll be off by more than twenty-four points because of statistical error; it happens, but not too often.

In the sixty-four-person poll, the curve has gotten thinner, so the margin of error has shrunk. Now, 95 percent of the time, the answer you get is somewhere between 61 percent and 39 percent for Bush—you're within plus or minus eleven points of the true answer.

The more people you ask, the smaller the margin of error. In the 1,024-person poll, the curve has gotten much thinner. The margin of error is now 3.1 percent; you're 95 percent confident that your poll is within 3.1 percentage points of the truth. The thinner the curve, the less likely it is that the poll gives an answer far away from the truth, and the smaller the margin of error.

As your sample gets larger and larger, the margin of error gets smaller and smaller. It never disappears completely, but it can be brought down to a level that's manageable—leaving the pollster to worry about other kinds of nonstatistical errors that can potentially render a poll meaningless.

Appendix B: Electronic Voting

Nowadays, no discussion of electoral fraud can be complete without at least a mention of electronic voting. A lot of critics, particularly on the left, are deeply concerned about the advent of electronic voting machines. They're terrified of the prospect of digital machines spiriting our votes away into the digital ether, perhaps allowing politicians to steal elections with a flick of a switch. It's true that electronic voting has the potential to cause electoral chaos. At the same time, though, it has great promise; it can make elections better, reducing errors and saving lots of time and money.

In the ideal—if the machines function perfectly—electronic voting has a lot of advantages. The most obvious is that people no longer have to count ballots; when a voter presses a button or a portion of the screen to indicate his preference, the vote is counted at the very same moment that it is cast. The act of voting is no longer separate from the act of tabulation. This alone saves a lot of time and money. Election officials no longer have to print (and store and transport) paper ballots, reducing costs even further. On top of that, machines are a boon to the disabled; they can prompt blind voters with audio signals, for example. It's no wonder that election officials want to switch over to electronic voting machines as soon as possible.

However, the machines have a downside. Unlike systems that use paper ballots, there's no physical record in a purely electronic voting machine. You press a button and you have to trust that the machine is recording your vote properly. You might never know if it gobbles up your vote or records it incorrectly. As a result, when things go wrong with the machines—as they occasionally do—they can completely obliterate people's votes. In the 2000 presidential election, a programming error wiped out at least 678 votes in New Mexico. (Al Gore "won" the state by a mere 366 votes.) Reports of vote-eating machines have surfaced all around the globe; in France, for example, a computer scientist reported in 2008 that precincts with electronic machines seemed to be swallowing votes at an alarming rate.* Also, there's the possibility that the machines will create votes out of thin air: in 2003, an election in Boone County, Iowa, initially registered 144,000 votes on electronic machines even though there were only 19,000 eligible voters. Sometimes machines rob one candidate and give votes to another. In 2000, one Florida county's electronic machines registered some *negative* 16,000 votes for Gore, while Bush received roughly 2,800 votes—all in a county with fewer than 600 registered voters. Many of the electronic voting machines on the market today are buggy and badly designed, so many voters are (justifiably) wary of trusting them at all.

How can engineers assure that the machines are trustworthy—

---

* In French districts with electronic voting machines there was nearly a 30 percent difference between the number of voters who turned up at polling places and the number of votes recorded, compared to about a 5 percent discrepancy in traditional paper-ballot districts.

that they're really counting our votes properly? It's a damnably hard problem. When you want to ensure that a transaction is recorded properly, you typically get a receipt; when you deposit a large amount of cash at a bank, for example, the ATM gives you a slip of paper that is proof the transaction took place. That way, if the bank fails to record the transaction, you can prove that they made an error. Unfortunately, you can't give receipts in a vote. If you did, it would open the door to some real nastiness: "Hey, buddy—if you don't give me a receipt proving that you voted for Gordon Brown in tomorrow's election, I'll break your kneecaps." The moment you give voters any form of proof that they voted for somebody, you create an environment that's ripe for voter intimidation or for vote selling. So you can't let us voters get our grubby paws on any sort of receipt.

Another possibility is that the machine prints out a receipt and lets the voter see it, but the voter can't touch it; instead, the receipt gets stored in the machine. This is a "voter-verified paper trail," and it has some advantages over purely electronic voting. However, it also has some serious disadvantages. For one thing, it's more costly, because you have to buy paper and ink to feed the machines. For another, paper jams and misfeeds will make the machines more likely to break down than a purely electronic machine. But these are minor issues. More serious is that the paper trails don't really solve the fundamental problem. Say that an election official notices that the machine counts are slightly different from what the paper trail shows. Do you believe the machine count or the paper trail? It's not at all clear. Sure, there might be a bug in the machine that created the discrepancy, but similarly, it might be that a few of the receipts got tangled up in the machinery or were lost by election workers or were simply miscounted (and as we saw, this often happens with paper ballots). Which is the source of the error? More often than not, machines are more reliable than humans, so it's far from certain that you should overrule the machine count with a human count of the receipts.

Voter-verified paper trails would increase the trustworthiness of electronic systems, but I think that a better solution would be to make the machines open-source. If the software is open to public scrutiny—if any programmer can search through the code looking for bugs—then we'd have a high degree of confidence that the machine is counting votes reliably. The same thing goes for the machine's hardware. So long as voting machines are proprietary, running buggy software that nobody can inspect, there is very good reason to be wary of trusting your votes to an electronic kiosk.

That being said, there has only been one reported incident of electoral fraud using electronic machines. In a 2007 Kentucky election, a judge, a court clerk, an elections official, and a bunch of other defendants allegedly schemed to fix local elections through a variety of means, including tampering with electronic voting machines. According to the indictment, the conspirators misled voters about how to use the machines. They told voters that their votes had been recorded when they pushed a button labeled "Vote" on the machine. In fact, there was an extra step—one that allowed the voter to review his vote and change it if desired. The vote wasn't officially cast until the voter pushed another button marked "Cast Ballot." So when voters walked out of the booth after having pressed "Vote," the conspirators allegedly waltzed in, changed the votes, and then cast the ballot once the votes had been "corrected." This scheme was low-tech. It exploited voter confusion rather than any inherent bug in the machine—and it shows that the threat can often come from a quarter you don't expect.

Electronic voting does indeed pose a potential risk to the reliability of our voting system, but that risk can be overcome. And if electronic voting is implemented properly, it can reduce the error rate inherent in elections; we no longer would be limited by humans' ability to count bits of paper accurately. All in all, electronic voting seems to be a fairly small problem—especially compared to the other ways that politicians use proofiness to game the electoral system.

Appendix C: The Prosecutor's Fallacy

The prosecutor's fallacy is a slimy little trick born out of the difficulty we humans have in interpreting probabilities. By failing to put a probability in the proper context, a lawyer can make it appear incontrovertible that a person is guilty when the case is far from certain.

To illustrate the fallacy, let's move away from the court system for a moment. Imagine that researchers have just discovered a rare and deadly disease—Head-Exploding Syndrome (HES). It's 100 percent fatal; if you contract it, you're going to die a horrible, painful, and very messy death. Luckily, researchers have developed an extremely good blood test to tell whether you've got HES. It's exquisitely accurate—there's only a one in a million chance that the blood test gives the wrong answer. Using this test, doctors start screening the population to find HES cases.

So you go in to the doctor's office, and the doctor draws your blood and leaves the room. A few minutes later, she returns, pale as a ghost. "The test came back positive," she says. Since the test is so accurate—the chances of an error are one in a million—it's virtually certain that you've got Head-Exploding Syndrome. There's only a one in a million chance that the test is wrong. This means that there's only a one in a million chance that you don't have the disease . . . right?

Not so fast. There's a piece of information that's missing before you can conclude that you've really got HES. You need to know just how rare the disease is. As it happens, HES is extraordinarily rare—epidemiologists estimate that it afflicts one in a billion people around the world. That means that of the seven billion people on earth, we expect only seven of them to have the disease. This little bit of information is crucial, because it allows you to put your positive test in context.

The one in a million chance of an error in the test seems pretty small, but if you're screening seven billion people, the one in a million chance of error means that roughly seven thousand patients are going to get a test result that gives the wrong answer. That is, seven thousand people around the planet will test positive on the test even though they don't have the disease. Since only seven people in the entire world actually have HES, this means that the vast majority of people who get a positive test don't actually have HES. Indeed, if you test positive, the probability that you have HES is 7 divided by 7,000—or 1 in 1,000. The chances are 999 in 1,000 that the test is wrong and you don't have the disease. You can rest easy. The one in a million probability of an incorrect test was deceptive; the probability didn't mean anything on its own. Only when you put that one in a million in context—comparing it to the one in a billion incidence of the disease—can you calculate your chance of actually having HES.

The lesson here is that you must always put probabilities in their proper context. It's a fallacy to look at the chance of the test's being wrong and equate that with the probability that you have a disease. Instead, you must compare the probability to the chances of having the disease in the first place—and when the disease is rare, it can make even a tiny probability of an error on a test loom large.

This is the prosecutor's fallacy in a different form. Instead of blood tests and disease, the fallacy deals with evidence and guilt, but the mathematics is exactly the same. A lawyer presents a very small probability without putting it in the proper context. As a result, the small probability convinces the jury that the statement must be true. Had the lawyer put the probability in the proper context, it would have been much less convincing—and perhaps even led the jury to the opposite conclusion.

For example, Alan Dershowitz argued that O. J. Simpson was innocent because there was only a one in a thousand chance that a wife-beater kills his wife. This number, he implies, means that there's only a one in a thousand chance that O. J. is guilty. This is the prosecutor's fallacy in action, because Dershowitz doesn't put the number into the proper context. The one in a thousand figure doesn't take into account that there is an extremely low probability that a thirty-five-year-old woman like Nicole Brown is murdered in a given year—about 1 in 40,000. Just as the one in a billion chance of disease made the one in a million chance of test error loom very large in comparison, the 1 in 40,000 chance of murder makes the 1 in 1,000 chance of a wife-beater turning murderer seem huge. When statisticians crunched the numbers in the proper way, their estimates of the probability of O. J.'s guilt turned out to be quite large—better than 50 percent. The fact that O. J. had previously battered his wife made it much more probable that he was the murderer, as would have been clear had Dershowitz put the number in the proper context.

Similarly, even if the probabilities Sir Roy Meadow used to convict Sally Clark of killing her children were in the ballpark (which they weren't), they would be deceptive on their own. Meadow didn't take into account that the probability of being a murderer—much less a serial murderer—in the United Kingdom is quite small. Any probability that Meadow used should have been put in the context of that very small probability, which would have made it look a lot less impressive to a jury. Indeed, a mathematician estimated the real probability to be closer to nine in ten that Sally Clark was not guilty based solely upon the deaths of her two children—a far, far cry from the 1 in 73 million chance that Meadow claimed.

The prosecutor's fallacy is powerful because it appeals to our innate misunderstanding of probabilities. It's counterintuitive that a tiny probability (a one in a million chance of a test's going wrong, for example) can, in certain contexts, wind up being extremely large. As a result, few people notice when a prosecutor ignores the context and, through a little numerical hanky-panky, makes a shaky case seem like a rock-solid one.

Notes

Introduction: Proofiness

1 **"In my opinion"** Joseph McCarthy quoted in Robert G. Torricelli and Andrew Carroll, *In Our Own Words* (New York: Washington Square Press, 1999), 174.

2 **"we have been able to compile"** "M'Carthy Insists Truman Oust Reds," *New York Times*, February 12, 1950, 5.

2n **"205 or 207?"** Quoted in David M. Oshinsky, *A Conspiracy So Immense* (New York: Free Press, 1983), 112.

3 **"Joe never had any names"** Ibid., 117.

1: Phony Facts, Phony Figures

7 The second chapter epigraph is drawn from Ronald Reagan, Address to the Republican National Convention, August 15, 1988. See, e.g., http://www.reagansheritage.org/html/reagan_rnc_88.shtml.

7 **"will become extinct by 2202"** "Blondes 'to die out in 200 years,'" BBC News, September 27, 2002, http://news.bbc.co.uk/2/hi/health/2284783.stm.

8 **"vanish from the face of the earth within two hundred years"** Quoted in Lawrence Altman, "Stop Those Presses! Blondes, It Seems, Will Survive After All," *New York Times*, October 2, 2002, A5.

8 **"WHO wishes to clarify"** "Clarification of erroneous news reports indicating WHO genetic research on hair color," World Health Organization, October 1, 2002, http://www.who.int/mediacentre/news/statements/statement05/en/.

8 **"Fifty-eight percent of all the exercise"** *Deborah Norville Tonight*, March 12, 2004, transcript at http://www.msnbc.msn.com/id/4533441/.

9n **"a five-year study"** *Countdown with Keith Olbermann*, September 8, 2004, transcript at http://www.msnbc.msn.com/id/5953239/.

13 **"40 per cent cleverer"** Denis Campbell, "Simple ways to make yourself far cleverer," *Observer* (UK), March 5, 2006.

14 **"twelve times more impact"** Television advertisement for L'Oreal Extra Volume Collagen Mascara, http://www.lorealparisusa.com/.

14 **"delivers 70 percent more moisture"** Television advertisement for Vaseline Dry Skin, http://www.vaseline.com/.

14 **29 percent of Christian Saints had exhibited psychic powers** Henry Reed and Charles Thomas Cayce, *Awakening Your Psychic Powers* (New York: Macmillan, 1988), 221.

15 **"From first puff to last"** Cigarette ad in *Life* magazine, October 24, 1955, available at http://tobaccodocuments.org/pollay_ads/Ches16.11.html.

16 **"If we say [the crowd] was 250,000"** Scott Bowles, "Park Police Can Count on a Disputed Crowd Figure; Aerial Photos, Bus Tallies Crucial to Accuracy," *Washington Post*, October 15, 1995.

17 **in fact, ABC had pegged the crowd** "ABC News Was Misquoted on Crowd Size," ABCNews.com, September 13, 2009, http://abcnews.go.com/Politics/protest-crowd-size-estimate-falsely-attributed-abc-news/story?id=8558055.

17 **the *Washington Post* estimated that the crowd was about 10,000** Philip Rucker, "Activists bring 'tea party' to Capitol Hill," *Washington Post*, November 6, 2009.

17 **Worse yet, some of *The Sean Hannity Show*'s footage** *The Daily Show with Jon Stewart*, November 10, 2009, http://www.thedailyshow.com/watch/tue-november-10-2009/sean-hannity-uses-glenn-beck-s-protest-footage.

17n **After the deception was exposed** Ibid., November 12, 2009, http://www.thedailyshow.com/watch/thu-november-12-2009/sean-hannity-apologizes-to-jon.

23 **The U.N insisted** "UN Chief welcomes six-billionth baby," BBC News, October 13, 1999, http://news.bbc.co.uk/2/hi/in_depth/472704.stm.

24 **the *Chicago Sun-Times* declared** "She's one in 300,000,000," *Chicago Sun-Times*, October 18, 2006, 6.

28 **"Five years ago"** George W. Bush, State of the Union Address, January 23, 2007.

29 **Twelfth-grade students' reading scores declined** NAEP Data Explorer, http://nces.ed.gov/nationsreportcard/naepdata/.

30 **"dramatic upward trend"** Jennifer Medina, "Reading and Math Scores Rise Sharply Across N.Y.," *New York Times*, June 24, 2008.

30 **"What a difference"** Michael Winerip, "One Secret to Better Test Scores: Make State Reading Tests Easier," *New York Times*, October 5, 2005.

33 **"Throughout the first 224 years"** "Bush Administration Breaks Record," press release from the Blue Dog Coalition, November 4, 2005.

33 **"$132 million, or 2.4 percent"** FYI: The American Institute of Physics Policy News, February 9, 2005, http://www.aip.org/fyi/2005/017.html.

35 **They emphasized the point with a graph** Ad available at http://cspinet.org/new/pdf/quaker.pdf.

38 **"The tax relief we passed"** George W. Bush, speech, February 19, 2004, http://georgewbush-whitehouse.archives.gov/news/releases/2004/02/20040219-4.html.

38 **most received less than $650** William G. Gale, Peter R. Orszag, and Isaac Sha-

piro, "Distributional Effects of the 2001 and 2003 Tax Cuts and Their Financing," Urban-Brookings Tax Policy Center, June 3, 2004, http://www.taxpolicycenter.org/UploadedPDF/411018_tax_cuts.pdf.

2: Rorschach's Demon

39 The chapter epigraph is drawn from Martin Gardner, *Fads and Fallacies in the Name of Science* (New York: Dover, 1952), 182.

39 **"Researchers Link Bad Debt to Bad Health"** *New York Times*, March 14, 2000, D8.

40n **"will continue with the Sisyphean task"** Ben Goldacre, "You are hereby sentenced eternally to wander the newspapers, fruitlessly mocking nutriwoo," *Guardian* (UK), July 19, 2008.

47 **"this carcinogenic substance"** Felicity Lawrence, "MP calls for ban on 'unsafe' sweetener," *Guardian* (UK), December 15, 2005.

48 **In 1984, there were 108 MRI machines** Earl P. Steinberg, "The Status of MRI in 1986: Rates of Adoption in the United States and Worldwide," *American Journal of Radiology*, September 1986, 453–455.

51 **"Marijuana and Your Teen's Mental Health"** Magazine advertisement, Office of National Drug Control Policy, ca. 2005.

52 **the "drug of entry into the world of hard drugs"** Adam Clymer, "Senate is told that cigarettes are 'entry' into hard drugs," *New York Times*, March 11, 1994, 22.

53n **"There is a mixed view among scientists and doctors"** Judy Keen, "Dole still 'not certain' if tobacco is addictive; supporters are baffled by his stance," *USA Today*, July 3, 1996, A5.

53n **"We know [smoking is] not good for kids"** "Clinton Assails Dole on Tobacco and Liquor Company on TV Ads," Associated Press, June 16, 1996.

56 **the "advantage" . . . evaporated with a new set of data** Analysis of Men's Taekwondo, Boxing, Freestyle Wrestling, and Greco-Roman Wrestling from the "Official Website of the Beijing 2008 Olympic Games," http://results.beijing2008.cn/WRM/ENG/Schedule/ and the NBC 2008 Beijing Olympic Games website http://www.2008.nbcolympics.com/.

64 **"The basic story from the equation"** Ray C. Fair, "The Effect of Economic Events on Votes for President: 1992 Update," *Political Behavior*, June 1996, 133.

64 **"It's not even going to be close"** Robert G. Kaiser, "Is This Any Way to Pick a Winner?" *Washington Post*, May 26, 2000, A1.

65 **the BBC trumpeted a formula** "The formula for happiness," BBC News, January 6, 2003, http://news.bbc.co.uk/2/hi/health/2630869.stm.

65n **I wrote an op-ed** Charles Seife, "Predicting the Presidents," *Washington Post*, June 3, 2000, A15.

66 **the formula that came from Cardiff University** "'I don't like Monday 24 January,'" BBC News, January 19, 2005, http://news.bbc.co.uk/2/hi/uk_news/4187183.stm.

66 **Such as the one described by this formula** Lukas I. Alpert, "Academics Develop Formula for Perfect Butt," *New York Post*, April 13, 2006.

3: Risky Business

67 The chapter epigraph is drawn from U.S. Congress, House of Representatives, Committee on Oversight and Government Reform, "Hearings on the Causes and Effects of the AIG Bailout," October 7, 2008.

67 **"Fate has ordained"** William Safire, memo to H. R. Haldeman, "In Event of Moon Disaster," July 18, 1969, available at http://www.thesmokinggun.com/archive/apollo1.html.

68 **less than 5 percent** Trudy E. Bell and Karl Esch, "The Space Shuttle: A Case of Subjective Engineering," *IEEE Spectrum*, June 1989, 42–46 (quotation on 44).

69 **"The probability of a . . . failure"** E. W. Colglazier and R. K. Weatherwax, "Failure Estimates for the Space Shuttle," Abstracts for Society for Risk Analysis, Annual Meeting 1986, quoted in Roger Cooke, *Experts in Uncertainty* (Oxford: Oxford University Press, 1991), 20.

69 **"to rely upon its engineering judgment"** Ibid.

70 **"As far as I can tell"** Richard Feynman, *What Do You Care What Other People Think?* (New York: Norton, 1988), 183.

70 **"It was clear"** Ibid.

71 **"Imagine that the U.S. is preparing"** Amos Tversky and Daniel Kahneman, "The Framing of Decisions and the Psychology of Choice," *Science*, January 30, 1981, 453.

73 **"Virgin has a detailed understanding"** virgingalactic.com, http://www.virgingalactic.com/htmlsite/faq.php?subtitle=Safety&src=141. (currently defunct)

74 **there would be roughly 275 U.S. plane crashes** Derived from Bureau of Transportation Statistics, "December 2008 Airline Traffic Data: System Traffic Down 5.7 Percent in December from 2007 and Down 3.7 Percent in 2008," http://www.bts.gov/press_releases/2009/bts012_09/html/bts012_09.html.

74 **New Mexico politicians** "Spaceport America Breaks Ground In New Mexico," press release, *spaceportamerica.com*, June 22, 2009, www.spaceportamerica.com/news/press-releases/244-spabreaksgroundnewmexico.html.

75 **"October 26, 2028 could be our last day"** Elizabeth Blue, *Herald* (Glasgow), March 13, 1998, 1.

75 **"The world ends on Feb 1 2019 (possibly)"** David Derbyshire, *Daily Telegraph* (UK), July 25, 2002, 3.

75 **"Is the end nigh? Science experiment could swallow Earth, critics say"** Joseph Brean, *National Post* (Canada), September 9, 2008, A1.

75 **"Physicists fear Big Bang machine could destroy Earth"** Jonathan Leake, *Calgary Herald*, July 19, 2000, A7.

76 **Audi sales plummeted** Marshall Schuon, "Audi Redesigns and Presses On," *New York Times*, April 18, 1993, section 8, p. 14.

77 **A safety "expert" had apparently bored a hole** Peter Huber, "Manufacturing the Audi Scare," *Wall Street Journal*, December 18, 1989, 1.

77n **"If that had happened at *60 Minutes*"** Elizabeth Kolbert, "NBC Settles Truck Crash Lawsuit, Saying Test Was 'Inappropriate,'" *New York Times*, February 10, 1993, A1.

78 **a few days before the war began** Philip Shenon and Eric Lichtblau, "U.S. Lowers Warning Level to 'Yellow,' but Cautions That Serious Threat Remains," *New York Times*, February 28, 2003, A14.

80n **"In 1985 Silverado's auditors"** William Sternberg, "Cooked Books," *Atlantic Monthly*, January 1992, 20.

87 **Paris's free-bicycle scheme was struggling to survive** "Thefts puncture Paris bike scheme," BBC News, February 10, 2009, http://news.bbc.co.uk/2/hi/europe/7881079.stm.

89 **"Nothing made me more frustrated"** "Bernanke on the Record," *NewsHour with Jim Lehrer*, July 26, 2009.

89 **"We are tightening our belt"** *NewsHour with Jim Lehrer*, October 22, 2008, transcript available at http://www.pbs.org/newshour/bb/business/july-dec08/aigcrisis_10-22.html.

4: Poll Cats

91 **"That is one for the books"** Caption for Byron Rollins photo entitled "Dewey Defeats Truman," November 4, 1948, www.apimages.com.

92 **"I think the 1948 polls"** Will Lester, "'Dewey Defeats Truman' Disaster Haunts Pollsters," Associated Press, November 1, 1998.

92n **"Jesus and Princess Diana"** Telegraph Online, June 8, 2009, http://www.telegraph.co.uk/news/newstopics/religion/5474600/Jesus-and-Princess-Diana-lead-poll-of-dead-people-we-most-want-to-meet.html.

94 **"It may be the act of going into bankruptcy"** Walter Lippmann, *Public Opinion*, part VII, chapter XXIII.

98n **"would win the White House"** Nitya Venkataraman, "Bush in Retrograde; Obama, McCain Rising," ABC News, June 2, 2008, http://abcnews.go.com/Politics/Vote2008/story?id=5286338&page=1.

98n **"It is cosmic writ"** "A Kerry win might be in the stars," Reuters, October 29, 2004.

104 **"Leader in every taste test"** Magazine advertisement for Heinz Tomato Juice, *Literary Digest*, November 14, 1936, 3.

104 **"This week, 500 pens"** "'The Digest' Presidential Poll is On!" *Literary Digest*, August 22, 1936, 3.

104 **a "king's ransom"** "'Digest' Poll Machinery Speeding Up," *Literary Digest*, August 29, 1936, 3.

105n **"a business proposition"** Quoted in George Gallup and Saul Forbes Rae, *The Pulse of Democracy* (New York: Simon and Schuster, 1940), 216.

106 **"The Poll represents"** "Landon, 1,293,669; Roosevelt, 972,897," *Literary Digest*, October 31, 1936, 4.

106 **"The mailing list is drawn from"** "'The Digest' Presidential Poll is On!"

107 **"Never before has anyone foretold"** Quoted in Gallup and Rae, *The Pulse of Democracy*, 47.

111 **"Average man sleeps with 7 women"** Associated Press, June 22, 2007.

113 **was around two hundred dollars** Nicole Wallace, "Donations to Hurricane-Relief Efforts Exceed $504-Million," *Chronicle of Philanthropy*, September 6, 2005.

114 **"seek out the exit poll takers"** Bill Carter, "3 Networks Admit Error in Arizona Primary Calls," *New York Times*, February 29, 1996, B9.

115 **"Hamas sought to influence the outcome"** Carl Bialik, "The Limits of Exit Polls," *Wall Street Journal*, February 21, 2006.

115 **in the early 1990s, Gallup . . . asked Americans** "Times Poll: Frequently Asked Questions," LATimes.com, http://www.latimes.com/news/custom/timespoll/la-timespollfaq,1,2370162.htmlstory.

118 **"a new Zogby poll"** "Zogby Poll: Americans Not in Favor of Starving Terri Schiavo," Zogby.com, April 1, 2005, http://www.zogby.com/Soundbites/readclips.cfm?ID=11131.

118 **"Do you agree or disagree"** "Zogby America Likely Voters 3/30/05 thru 4/2/05 MOE +/-3.2 percentage points," Zogby.com, April 2, 2005, http://operationrescue.org/files/wf-Schiavo.pdf.

121 **"AP Poll: Americans optimistic for 2007"** Nancy Benac, "AP Poll: Americans optimistic for 2007," Associated Press, December 30, 2006.

121 **"Poll: Americans see doom, gloom for 2007"** Darlene Superville, "Poll: Americans see doom, gloom for 2007," Associated Press, December 31, 2006.

122 **"If you're disgusted with us"** See, e.g., "Election Night: A Media Watch Special Report," PBS.org, http://www.pbs.org/newshour/media/election2000/election_night.html, or "Dan Rather Opines on Politics and the Media," *Larry King Live*, January 10, 2001, http://transcripts.cnn.com/TRANSCRIPTS/0101/10/lkl.00.html.

122 **"Sip it, savor it"** See, e.g., Alexander Stille, "How the Press Saw the 2000 Election," *New York Times*, January 8, 2003.

5: Electile Dysfunction

127 **"If a voter has written"** Minnesota Election Laws 2008, Ch.204C.22 subd. 4.

127 **"If it said 'Moon Unit Zappa'"** Minnesota State Canvassing Board meeting, December 18, 2008.

131 **"How could anyone bribe someone with Twinkies?"** "'Twinkiegate' Candidate Faces Charges," Associated Press, May 23, 1986.

131 **The case ended a few weeks later** "Twinkiegate Case Closed," *Houston Chronicle*, June 5, 1986, 11.

132 **"Yesterday, the voters spoke"** Jason Hoppin, Dave Orrick, and Rachel Stassen-Berger, "Recount in Senate race evokes memories of 2000," *St. Paul Pioneer Press*, November 6, 2008.

132 **"If you ask me what I would do"** Norm Coleman, press conference, November 5, 2008, available at http://www.talkingpointsmemo.com/archives/243080.php.

133 **a clerk had accidentally typed** Patricia Lopez, "Franken's Deficit: 236 Votes," *Minneapolis Star-Tribune*, November 7, 2008.

134 **"[steal] a Senate seat for left-wing joker Al Franken"** "Mischief in Minnesota," *Wall Street Journal*, November 12, 2008.

134 **"statistically dubious"** Josh Kraushaar, "No joke: Minnesota race is sooooo ugly . . . ," Politico.com, November 11, 2008, http://www.politico.com/news/stories/1108/15490.html.

134n **driving around with them in the back of her car** David Brauer, "Minneapolis election director speaks: 'Ballots in my car' story false," MinnPost.com, http://www.minnpost.com/davidbrauer/2008/11/12/4565/minneapolis_election_director_speaks_ballots_in_my_car_story_false.

135 **"cut both ways"** "Mischief in Minnesota."

137 **Minneapolis's ABC affiliate commissioned a poll** Michelle Knoll, "Poll: Senate results unchanged if election held today," December 8, 2008, http://kstp.com/article/stories/S695337.shtml.

138 **the engineer was surprised to discover** Charles King Emma, "Observations on Physical Inventory and Stock Record Error," *Journal of Industrial Engineering*, July 1967, 396–402.

138 **"Humans counting things"** "Every Vote Counts, Unless You Live in MN," *Talk of the Nation*, December 8, 2008, http://www.npr.org/templates/player/mediaPlayer.html?action=1&t=1&islist=false&id=97960769&m=97960763.

140 **"Our accounting procedures dictate"** Charles Seife, telephone conversation with Fran Windschitl, November 29, 2008.

140 **In a precinct in Blue Earth County** Charles Seife, calculation based upon election data from the Minnesota Secretary of State's office, http://www.sos.state.mn.us/home/index.asp.

143 **"I've been before the state canvassing board"** Charles Seife, firsthand observation of Ramsey County recount, November 19, 2008.

145 **"The Republican incumbent held a slight edge"** Brian Bakst, "Coleman keeps edge as hand recount wraps up," Associated Press, December 5, 2009.

145 **Nate Silver used regression analyses** Nate Silver, "Projection: Franken to win by 27 votes," FiveThirtyEight.com, November 23, 2008, http://www.fivethirtyeight.com/2008/11/projection-franken-to-win-recount-by-27.html.

145 **Two weeks later, he reversed himself** Nate Silver, "Statistical Models Now Show Coleman as Slight Favorite," FiveThirtyEight.com, December 2, 2008, http://www.fivethirtyeight.com/2008/12/statistical-models-now-show-coleman-as.html.

148 **"There is no more precious right in a democracy"** Minnesota State Canvassing Board meeting, November 18, 2008.

149 **"It's actually the responsibility of the voter"** Fritz Knaak, impromptu conversation with the press in the Minnesota statehouse, November 18, 2008.

151 **"were cast in areas which favored Coleman"** *Coleman et al. v. Franken*, No. 62-CV-09-56, Contestee's Answer and Counterclaims, January 12, 2009, 7.

153 **"error in the vote tabulation"** 2000 Florida Statutes, Title IX, Ch. 102, Part 166, §(4)(c).

154n **"Palm Beach County is a Pat Buchanan stronghold"** Jake Tapper, "Buchanan camp: Bush claims are 'nonsense,'" Salon.com, November 10, 2000, http://dir.salon.com/story/politics/feature/2000/11/10/buchanan/.

157 **"I can not believe"** Mark Whitman, ed., *Florida 2000: A Sourcebook on the Contested Presidential Election* (Boulder, CO: Lynne Rienner Publishers, 2003), 106.

159 **"It is obvious"** *Bush v. Gore*, majority opinion, December 12, 2000.

160n "Our consideration is limited" Ibid.

160n "Respectfully, the Supreme Court" *Stewart v. Blackwell*, majority opinion, April 21, 2006.

162 "In twenty-five precincts" *The Journal Editorial Report*, Fox News, January 10, 2009.

162 closer to seven hundred precincts Charles Seife, calculation based upon election data from the Minnesota Secretary of State's office, http://www.sos.state.mn.us/home/index.asp.

163 "brought into sharp focus" *Bush v. Gore*, majority opinion, December 12, 2000.

6: An Unfair Vote

181n "Project Ratfuck" Jeffrey Toobin, "The Great Election Grab," *New Yorker*, December 8, 2003.

182 the Republicans called in the Department of Homeland Security Henry A. Waxman et al., letter to Tom Ridge, June 4, 2003.

188 all but about 2 percent of the population Department of Commerce, *U.S. Census Monitoring Board, Final Report to Congress*, September 1, 2001, 6.

192 "Properly designed sampling" *Department of Commerce v. U.S. House of Representatives*, Amicus Curiae Brief of the American Statistical Association.

193 will "give the party controlling Congress" *Department of Commerce v. U.S. House of Representatives*, Scalia's concurrence.

193 It was inserted during the drafting of the U.S. Constitution Margaret Mikyung Lee, "Sampling for Census 2000: A Legal Overview," Library of Congress, Congressional Research Service, Report CRS-97-871, updated December 30, 1998, 3.

193n One theory is that people were afraid Ibid.

196 "well familiar with methods of estimation" *Utah v. Evans*, 536 U.S. 452 (2002), Thomas's opinion.

7: Alternate Realities

199 The chapter epigraph is drawn from Angela Tuck, "'Objective truth' passes the test for letters to the editor," *Atlanta Journal-Constitution*, August 4, 2007, A15.

200 "wholly wanting and misleading" Will E. Edington, "House Bill No. 246, Indiana State Legislature, 1897," *Proceedings of the Indiana Academy of Science* 45 (1935): 207.

200 The bill got off to a great start "Indiana pi, " http://www.agecon.purdue.edu/crd/Localgov/Second%20Level%20pages/Indiana_Pi_Story.htm.

200 "might as well try to legislate" Edington, "House Bill No. 246, Indiana State Legislature, 1897," 210.

202 ". . . 50 percent—perhaps as high as 80 percent" I. G. Good, "When batterer turns murderer," *Nature*, June 15, 1995, 541; John Allen Paulos, "Murder he wrote," http://www.math.temple.edu/~paulos/oj.html.

203 He was so skeptical of the condition Roy Meadow, "Unnatural Sudden Infant Death," *Archives of Disease in Childhood* 80 (1999): 7–14.

204 A blistering article Stephen J. Watkins, "Conviction by mathematical error?" *British Medical Journal*, January 1, 2000, 2–3.

204 **"invalid"** Peter Green, "Letter from the President to the Lord Chancellor regarding the use of statistical evidence in court cases," January 23, 2002.
205 **only if a jury finds that he would continue** Texas Code of Criminal Procedure, Article 37.071.
205 **Grigson was nicknamed "Dr. Death"** "They Call Him Dr. Death," *Time*, June 1, 1981.
206 **"The large body of research"** *Barefoot v. Estelle*, 463 U.S. 880 (1983), Amicus Curiae Brief of the American Psychiatric Association, October 1982.
207 **"A layman with access"** *Barefoot v. Estelle*, dissent.
207 **"The suggestion that no psychiatrist's testimony"** *Barefoot v. Estelle*, majority opinion.
207 **"Sharon, tell all my friends goodbye"** Texas Department of Criminal Justice, "Last Statement—Thomas Barefoot," http://www.tdcj.state.tx.us/stat/barefoot thomaslast.htm.
208 **By the mid-1980s** "They Call Him Dr. Death." *Time*, June 1, 1981.
208 **expelled Grigson for ethics violations** Laura Beil, "Groups Expel Texas Psychiatrist Known for Murder Cases," *Dallas Morning News*, July 26, 1995, A36.
209 **"an increase in traffic of at least 100 percent"** 49 CFR 1105.7(e)(5)(i)(A).
209 **"The cities believe"** *Lee's Summit v. Surface Transportation Board*, 231 F.3d 39; 343 U.S. App. D.C. 405, Decision, November 14, 2000.
211 **"The result of these blood grouping tests"** *Berry v. Chaplin*, 74 Cal. App. 2d 652; 169 P.2d 442; May 27, 1946.
212 **"The DNA test Osborne seeks"** *District Attorney's Office for the Third Judicial District v. Osborne*, 521 F. 3d 1118 (2009), dissent.
214 **"to prevail under the Equal Protection Clause"** *McCleskey v. Kemp*, 481 U.S. 279 (1987), majority opinion.
215 **"McCleskey challenges decisions"** Ibid.
215 **"Since it is my view"** Quoted in Dennis D. Dorin, "Far Right of the Mainstream: Racism, Rights, and Remedies from the Perspective of Justice Antonin Scalia's *McCleskey* Memorandum," *Mercer Law Review* 45 (1993–1994): 1038.
216 **"Yes, *McCleskey v. Kemp*"** John Calvin Jeffries, *Justice Lewis F. Powell Jr.* (New York: Fordham University Press, 2001), 451.
216 **"a pattern of evidence"** General Accounting Office, "Death Penalty Sentencing: Research Indicates Pattern of Racial Disparities," GAO-GGD-90-57, February 1990, 5.
217 **"The legitimacy of deterrence"** *Baze v. Rees*, 553 U.S. ____ (2008), Stevens concurrence.
217 **"Justice Stevens' analysis"** *Baze v. Rees*, Scalia concurrence.
218 **"In short, the best reading"** Cass Sunstein and Justin Wolfers, "A Death Penalty Puzzle," *Washington Post*, June 30, 2008, A11.
219 **"remarkable in number"** *Kansas v. Marsh*, dissent, June 26, 2006.
219 **listed 340 prisoners** Samuel R. Gross et al., "Exonerations in the United States 1989 through 2003," *Journal of Criminal Law and Criminology* 95, no. 2 (2005): 523–560.
219 **"So, let's give the professor"** Joshua Marquis, "The Innocent and the Shammed," *New York Times*, January 26, 2006, A23.

220 **Rapes and murders are just a tiny percentage** Matthew R. Durose and Patrick A. Langan, "State Court Sentencing of Convicted Felons, 2002 Statistical Tables," NCJ 208910, Department of Justice, Office of Justice Programs, Bureau of Justice Statistics, May 2005.

221 **between 3 and 5 percent of those convictions** D. Michael Risinger, "Innocents Convicted: An Empirically Justified Factual Wrongful Conviction Rate," *Journal of Criminal Law and Criminology* 97, no. 3 (2007): 761–806.

8: Propaganda by the Numbers

223 The chapter epigraph is drawn from Jacques Ellul, *Propaganda: The Formation of Men's Attitudes*, translated by Konrad Kellen and Jean Lerner (New York: Vintage, 1965), 85–86.

225 **"Time and again"** "Briefings: A Ritual of Noncommunication," *Time*, October 10, 1969.

225 **"encouraged"** Charles Seife, "Pie in the sky," *New Scientist*, April 10, 1999, 1414.

226 **"I was trained to believe"** Samuel Freedman, *Letters to a Young Journalist* (New York: Basic Books, 2006), 50.

227 **"One would like to cite countless facts"** Joseph Alsop, "Joint Vietnamization Efforts Are Working in Delta Hamlets," *Washington Post*, September 26, 1969, A27.

228 **was better for the environment** CNW Marketing Research, Inc., "Dust to Dust: The Energy Cost of New Vehicles From Concept to Disposal," March 2007, available at http://cnwmr.com/nss-folder/automotiveenergy/.

228n **"In the political backrooms"** Joseph C. Harsch, "The Politics of Peace," *Christian Science Monitor*, January 9, 1969, 18.

229 **"Speaking of Hummers"** George F. Will, "Fuzzy Climate Math," *Washington Post*, April 12, 2007, A27.

230 **"the possibility of increased risk of breast cancer"** Texas Health and Safety Code §171.012(1)(B)(iii).

230 **a 30 percent greater chance** J. Brind et al., "Induced abortion as an independent risk factor for breast cancer: a comprehensive review and meta-analysis," *Journal of Epidemiology and Community Health* 50 (1996): 481–496.

232 **"increased risk of suicidal ideation and suicide"** South Dakota Codified Laws, 34-23A-10.1(1)(e)(ii).

232 **According to a 2005 Finnish study** Mika Gissler et al., "Injury deaths, suicides, and homicides associated with pregnancy, Finland, 1987–2000," *European Journal of Public Health* 15, no. 5 (2005): 459–463.

233 **a great deal of praise and attention** James Lindgren, "Fall from Grace: *Arming America* and the Bellesiles Scandal," *Yale Law Journal*, June 2002, 2195–2249.

233 **one set of Bellesiles's numbers was mathematically impossible** Ibid., 2198.

233 **"extremely sloppy documentation"** Stanley N. Katz, Hanna H. Gray, and Laurel Thatcher Ulrich, "Report of the Investigative Committee in the matter of Professor Michael Bellesiles," Emory University, July 10, 2002, 11, available at http://www.emory.edu/news/Releases/Final_Report.pdf.

233 **"evidence of falsification"** Ibid., 19.

237 **"On several occasions seismic signals"** Judith Miller, "U.S. Confirms a Plan to Halt Talks on a Nuclear Test Ban," *New York Times*, July 21, 1982, A1.

238 **"had been estimated at 300 kilotons"** Judith Miller, "Experts Split on Flaws in Pacts Limiting Nuclear Tests," *New York Times*, July 26, 1982, A3.

238 **they came up with an excellent equation** Gregory E. van der Vink, "The Role of Seismologists in Debates over the Comprehensive Test Ban Treaty," *Annals of the New York Academy of Sciences* 866 (1998): 84–113.

239 **"People started circulating calculations of yield"** Charles Seife, telephone communication with Fred Lamb, July 21, 2009.

240 **"[He] actually accused"** Ibid.

240 **"Soviet nuclear testing activities"** U.S. Department of State Bulletin, February 1986, available at http://findarticles.com/p/articles/mi_m1079/is_v86/ai_4115455/.

240 **"As evidence continued to accumulate"** Charles Seife, telephone communication with Paul Richards, August 5, 2009.

Appendix B: Electronic Voting

252 **a programming error wiped out** Dan Keating, "Lost Votes in N.M. a Cautionary Tale," *Washington Post*, August 22, 2004, A5.

252 **in France, . . . a computer scientist reported** Peter Sayer, "Study: Electronic Voting Increased Counting Errors in France," PCWorld.com, July 9, 2008, http://www.pcworld.com/printable/article/id,148112/printable.html.

254 **In a 2007 Kentucky election** *United States of America v. Russell Cletus Maricle et al.*, No. 09-16-S-DCR. U.S. District Court, Eastern District of Kentucky, Southern Division, London, Indictment, July 9, 2009.

Bibliography

"ABC News Was Misquoted on Crowd Size." ABCNews.com, September 13, 2009. http://abcnews.go.com/Politics/protest-crowd-size-estimate-falsely-attributed-abc-news/story?id=8558055.

Agresti, Alan, and Brett Parnell. "Misvotes, Undervotes and Overvotes: The 2000 Presidential Election in Florida." *Statistical Science* 17, no. 4 (2002): 436–440.

Alexander, Michele G., and Terri D. Fisher. "Truth and Consequences: Using the Bogus Pipeline to Examine Sex Differences in Self-Reported Sexuality." *Journal of Sex Research*, February 2003, 27–35.

Alpert, Lukas I. "Academics Develop Formula for Perfect Butt." *New York Post*, April 13, 2006.

Alsop, Joseph. "Joint Vietnamization Efforts Are Working in Delta Hamlets." *Washington Post*, September 26, 1969, A27.

Altman, Lawrence. "Stop Those Presses! Blondes, It Seems, Will Survive After All." *New York Times*, October 2, 2002, A5.

American Cancer Society. "Can Having an Abortion Cause or Contribute to Breast Cancer?" http://www.cancer.org/docroot/CRI/content/CRI_2_6x_Can_Having_an_Abortion_Cause_or_Contribute_to_Breast_Cancer.asp.

"American Experience: LBJ." PBS. Transcript available at http://www.pbs.org/wgbh/amex/presidents/36_l_johnson/filmmore/filmscript.html.

Angier, Natalie. "2 Experts Say Women Who Run May Overtake Men." *New York Times*, January 7, 1992.

Ansolabehere, Stephen, and Andrew Reeves. "Using Recounts to Measure the Accuracy of Vote Tabulations: Evidence from New Hampshire Elections 1946–2002." *Caltech/MIT Voting Technology Project*, VTP Working Paper #11, January 2004.

"Average man sleeps with 7 women." Associated Press, June 22, 2007.

Avery v. Georgia, 345 U.S. 559 (1953).

Aynesworth, Hugh. "Texas 'Dr. Death' retires after 167 capital case trials; Felt murderers would kill again." *Washington Times*, December 21, 2003, A02.

Backstrom, Charles H., and Leonard Robins. "The Supreme Court Prohibits Gerrymandering: A Gain or a Loss for the States?" *Publius*, Summer 1987, 101–109.

Baker v. Carr, 369 U.S. 186 (1962).

Bakst, Brian. "Coleman keeps edge as hand recount wraps up." Associated Press, December 5, 2009.

———. "Most 'undervotes' cast in counties won by Obama." Associated Press, November 7, 2008.

Baldus, David C., Charles Pulaski, and George Woodworth. "Comparative Review of Death Sentences: An Empirical Study of the Georgia Experience." *Journal of Criminal Law and Criminology*, Autumn 1983, 661–753.

Baldus, David C., et al. "Identifying Comparatively Excessive Sentences of Death: A Quantitative Approach." *Stanford Law Review*, November 1980, 1–74.

Barefoot v. Estelle, 463 U.S. 880 (1983).

Bartlett v. Strickland, 556 U.S. ____ (2009).

Baze v. Rees, 553 U.S. ____ (2008).

Beil, Laura. "Groups Expel Texas Psychiatrist Known for Murder Cases." *Dallas Morning News*, July 26, 1995, A36.

Bell, Trudy E., and Karl Esch. "The Space Shuttle: A Case of Subjective Engineering." *IEEE Spectrum*, June 1989, 42–46.

Bellesiles, Michael A. "The Origins of Gun Culture in the United States, 1760–1865." *Journal of American History*, September 1996, 425–455.

———. "Statement of Michael Bellesiles on Emory University's Inquiry into Arming America." Ca. July 2002. Available at http://www.emory.edu/news/Releases/B_statement.pdf

Benac, Nancy. "AP Poll: Americans optimistic for 2007." Associated Press, December 30, 2006.

Benjamin, Burton. *The CBS Benjamin Report*. Washington, D.C.: Media Institute, 1984.

"Bernanke on the Record." *NewsHour with Jim Lehrer*, July 26, 2009.

Berry v. Chaplin, 74 Cal. App. 2d 652; 169 P.2d 442 (1946).

Berry v. Chaplin, 74 Cal. App. 2d 669; 169 P.2d 453 (1946).

Best, Joel. *Damned Lies and Statistics*. Berkeley: University of California Press, 2001.

———. *More Damned Lies and Statistics*. Berkeley: University of California Press, 2004.

Bialik, Carl. "The Limits of Exit Polls." *Wall Street Journal*, February 21, 2006.

Blau, Julian H. "A Direct Proof of Arrow's Theorem." *Econometrica*, January 1972, 61–67.

Blaze, Matt. "Is the e-voting honeymoon over?" March 23, 2009. http://www.crypto.com/blog/vote_fraud_in_kentucky/.

"Blondes 'to die out in 200 years.'" BBC News, September 27, 2002. http://news.bbc.co.uk/2/hi/health/2284783.stm.

Blue Dog Coalition. "Bush Administration Breaks Record." Press release, November 4, 2005.

Blue, Elizabeth. "October 26, 2028 could be our last day." *Herald* (Glasgow), March 13, 1998, 1.

Boorstin, Daniel J. *The Image: A Guide to Pseudo-Events in America*. New York: Vintage, 1987.

Bowen, Debra. "California Secretary of State Debra Bowen's Report to the Election Assistance Commission Concerning Errors and Deficiencies in Diebold/Premier GEMS Version 1.18.19." March 2, 2009. Available at www.sos.ca.gov/elections/voting_systems/sos-humboldt-report-to-eac-03-02-09.pdf.

Bowles, Scott. "Park Police Can Count on a Disputed Crowd Figure; Aerial Photos, Bus Tallies Crucial to Accuracy." *Washington Post*, October 15, 1995.

Bradlee, Benjamin C. *Conversations with Kennedy*. New York: Norton, 1975.

Brauer, David. "Minneapolis election director speaks: 'Ballots in my car' story false." MinnPost.com, November 12, 2008. http://www.minnpost.com/davidbrauer/2008/11/12/4565/minneapolis_election_director_speaks_ballots_in_my_car_story_false.

Brean, Joseph. "Is the end nigh? Science experiment could swallow Earth, critics say." *National Post*, September 9, 2008, A1.

"Briefings: A Ritual of Noncommunication." *Time*, October 10, 1969.

Brigham, Erwin R. "Pacification Measurement." *Military Review*, May 1970, 47–55.

Brind, J., et al. "Induced abortion as an independent risk factor for breast cancer: a comprehensive review and meta-analysis." *Journal of Epidemiology and Community Health* 50 (1996): 481–496.

Brown, Curt. "Minnesota's vote: Cast into doubt." *Minneapolis Star-Tribune*, December 14, 2008.

Bryson, Maurice C. "The Literary Digest Poll: Making of a Statistical Myth." *American Statistician*, November 1976, 184–185.

Bureau of Transportation Statistics. "December 2008 Airline Traffic Data: System Traffic Down 5.7 Percent in December from 2007 and Down 3.7 Percent in 2008." http://www.bts.gov/press_releases/2009/bts012_09/html/bts012_09.html.

Bush, George W. "Address to the Nation on Iraq," March 17, 2003. Available at http://www.gpo.gov/fdsys/pkg/WCPD-2003-03-24/html/WCPD-2003-03-24-Pg338-2.htm.

———. *Homeland Security Presidential Directive 3*. "Subject: Homeland Security Advisory System," March 11, 2002.

———. Speech, February 19, 2004. http://georgewbush-whitehouse.archives.gov/news/releases/2004/02/20040219-4.html.

———. State of the Union Address, January 23, 2007.

Bush v. Gore, 531 U.S. 98 (2000).

Bush v. Palm Beach County Canvassing Board, 531 U.S. 70 (2000).

Bush v. Palm Beach County Canvassing Board, 531 U.S. 1050 (2000).

Butler v. Thompson, 97 F. Supp. 17 U.S. District Court for the Eastern District of Virginia (1951).

Callins v. Collins, 510 U.S. 1141 (1994).

Campbell, Denis. "Simple ways to make yourself far cleverer." *Observer* (UK), March 5, 2006.

Caption for Byron Rollins photo entitled "Dewey Defeats Truman." Associated Press, November 4, 1948. www.apimages.com.

Carter, Bill. "3 Networks Admit Error in Arizona Primary Calls." *New York Times*, February 29, 1996, B9.

Center for Science in the Public Interest. "Old Quaker Labels and Ads." http://cspinet.org/new/pdf/quaker.pdf.

Cigarette ad in *Life* magazine, October 24, 1955. http://tobaccodocuments.org/pollay_ads/Ches16.11.html.

"Clarification of erroneous news reports indicating WHO genetic research on hair color." World Health Organization, October 1, 2002. http://www.who.int/mediacentre/news/statements/statement05/en/.

Clark, R v. [2000] EWCA Crim 54 (October, 2, 2000).

"Clinton Assails Dole on Tobacco and Liquor Company on TV Ads." Associated Press, June 16, 1996.

Clymer, Adam. "Senate is told that cigarettes are 'entry' into hard drugs." *New York Times*, March 11, 1994, 22.

CNW Marketing Research, Inc. "Dust to Dust: The Energy Cost of New Vehicles from Concept to Disposal," March 2007. Available at http://cnwmr.com/nss-folder/automotiveenergy/.

Coleman, Norm. Press conference, November 5, 2008. Available at http://www.talkingpointsmemo.com/archives/243080.php.

Coleman et al. v. Franken, No. 62-CV-09-56 (2009).

Cooke, Roger. *Experts in Uncertainty*. Oxford: Oxford University Press, 1991.

Countdown with Keith Olbermann, September 8, 2004. Transcript at http://www.msnbc.msn.com/id/5953239/.

County of Stearns, Minnesota. "U.S. Senate Race Recount Facing More Challenges." Press release, November 24, 2008.

Cramer, Clayton E. "Why Footnotes Matter: Checking *Arming America*'s Claims." *Plagiary: Cross-Disciplinary Studies in Plagiarism, Fabrication, and Falsification* 1, no. 11 (2006): 1–31.

Croen, Lisa A., et al. "The Changing Prevalence of Autism in California." *Journal of Autism and Developmental Disorders*, June 2002, 207–215.

The Daily Show with Jon Stewart, November 10, 2009. http://www.thedailyshow.com/watch/tue-november-10-2009/sean-hannity-uses-glenn-beck-s-protest-footage.

——. November 12, 2009. http://www.thedailyshow.com/watch/thu-november-12-2009/sean-hannity-apologizes-to-jon.

Daly, Christopher, and Hamil Harris. "Boston U. Sets March At 837,000." *Washington Post*, October 28, 1995, C3.

"Dan Rather Opines on Politics and the Media." *Larry King Live*, January 10, 2001. http://transcripts.cnn.com/TRANSCRIPTS/0101/10/lkl.00.html.

Davis, Don. "Senate Race in Overtime." *Worthington (MN) Daily Globe*, November 6, 2008.

Davis v. Bandemer, 478 U.S. 109 (1986).

Deborah Norville Tonight, March 12, 2004. Transcript at http://www.msnbc.msn.com/id/4533441/.

Department of Commerce. *U.S. Census Monitoring Board, Final Report to Congress*, September 1, 2001, 6.

Department of Commerce v. U.S. House of Representatives, 525 U.S. 326.

Derbyshire, David. "The world ends on Feb 1 2019 (possibly)." *Daily Telegraph* (UK), July 25, 2002, 3.

de Sola Pool, Ithiel, et al. *Hamlet Evaluation System Study (HES) ACG60F.* Cambridge, MA: Simulmatics Corporation, May 1, 1968.

Diaz, Kevin, and Bob von Sternberg. "$12 million later, still no senator from Minnesota." *Minneapolis Star-Tribune*, April 16, 2009.

Donohue, John J., and Justin Wolfers. "Uses and Abuses of Empirical Evidence in the Death Penalty Debate." *Stanford Law Review* 58, no. 3 (2006): 791–845.

Dorin, Dennis D. "Far Right of the Mainstream: Racism, Rights, and Remedies from the Perspective of Justice Antonin Scalia's *McCleskey* Memorandum." *Mercer Law Review* 45 (1993–1994): 1036–1088.

"'Digest' Poll Machinery Speeding Up." *Literary Digest*, August 29, 1936, 3.

"'The Digest' Presidential Poll is On!" *Literary Digest*, August 22, 1936, 3.

District Attorney's Office for the Third Judicial District v. Osborne, 521 F. 3d 1118 (2009).

Doyle, Pat, and Mike Kaszuba. "Coleman makes dramatic switch in his strategy." *Minneapolis Star-Tribune*, January 19, 2009.

Drentea, Patricia, and Paul J. Lavrakas. "Over the limit: the association among health, race, and debt." *Social Science and Medicine* 50 (2000): 517–529.

Duchschere, Kevin, Curt Brown, and Pam Louwagie. "Recount: The Coleman-Franken brawl drags on." *Minneapolis Star-Tribune*, November 5, 2008.

Duchschere, Kevin, and Mike Kaszuba. "Senate race certification set for Monday with Franken up 225 votes." *Minneapolis Star-Tribune*, January 4, 2009.

Durose, Matthew R., and Patrick A. Langan. "State Court Sentencing of Convicted Felons, 2002 Statistical Tables." NCJ 208910. Department of Justice, Office of Justice Programs, Bureau of Justice Statistics, May 2005.

Dyer, Claire. "Sally Clark freed after appeal court quashes her convictions." *British Medical Journal*, February 8, 2003, 304.

Edington, Will E. "House Bill No. 246, Indiana State Legislature, 1897." *Proceedings of the Indiana Academy of Science* 45 (1935): 206–210.

Eggen, Dan. "Justice Staff Saw Texas Districting as Illegal." *Washington Post*, December 2, 2005, A1.

"Election Night: A Media Watch Special Report." PBS.org. http://www.pbs.org/newshour/media/election2000/election_night.html.

Ellul, Jacques. *Propaganda: The Formation of Men's Attitudes.* Translated by Konrad Kellen and Jean Lerner. New York: Vintage, 1965.

Emma, Charles King. "Observations on Physical Inventory and Stock Record Error." *Journal of Industrial Engineering*, July 1967, 396–402.

Enguehard, Chantal. "Transparency in Electronic Voting: the Great Challenge." Paper presented at IPSA International Political Science Association RC 10 on Electronic Democracy, conference on "E-democracy—State of the art and future agenda, " Stellenbosch University, South Africa, January 22–24, 2008.

"Every Vote Counts, Unless You Live in MN. *Talk of the Nation*, December 8, 2008. http://www.npr.org/templates/player/mediaPlayer.html?action=1&t=1&islist=false&id=97960769&m=97960763.

Fair, Ray C. "The Effect of Economic Events on Votes for President." *Review of Economics and Statistics*, May 1978, 159–173.

———. "The Effect of Economic Events on Votes for President: 1988 Update." http://fairmodel.econ.yale.edu/rayfair/pdf/1990C200.pdf.

———. "The Effect of Economic Events on Votes for President: 1992 Update." *Political Behavior*, June 1996, 119–139.

"Farewell to the Follies." *Time*, February 12, 1973.

Feige, David. "With Math Skills Like These, It's No Wonder Scalia is a Lawyer." Slate.com, March 27, 2008. http://www.slate.com/blogs/blogs/convictions/archive/2008/03/27/with-math-skills-like-these-it-s-no-wonder-scalia-is-a-lawyer.aspx.

Feynman, Richard. *What Do You Care What Other People Think?* New York: Norton, 1988.

Fladell v. Palm Beach County Canvassing Board, SC00-2373, SC00-2376 (2000).

Florida Commission on Capital Cases. "Case Histories: A Review of 24 Individuals Released from Death Row," September 10, 2002. http://www.floridacapitalcases.state.fl.us/Publications/innocentsproject.pdf.

Florida Statutes (2000).

"The formula for happiness." BBC News, January 6, 2003. http://news.bbc.co.uk/2/hi/health/2630869.stm.

Freedman, Samuel. *Letters to a Young Journalist*. New York: Basic Books, 2006.

Fried, Richard M. *Men Against McCarthy*. New York: Columbia University Press, 1976.

Fryar, Cheryl, et al. "Drug Use and Sexual Behaviors Reported by Adults: United States, 1999–2002." Advance Data from Vital and Health Statistics, no. 384. National Center for Health Statistics.

FYI: The American Institute of Physics Policy News, February 9, 2005. http://www.aip.org/fyi/2005/017.html.

Gale, William G., Peter R. Orszag, and Isaac Shapiro. "Distributional Effects of the 2001 and 2003 Tax Cuts and Their Financing." Urban-Brookings Tax Policy Center, June 3, 2004. http://www.taxpolicycenter.org/UploadedPDF/411018_tax_cuts.pdf.

Gallup, George, and Saul Forbes Rae. *The Pulse of Democracy*. New York: Simon and Schuster, 1940.

Gardner, Martin. *Fads and Fallacies in the Name of Science*. New York: Dover, 1952.

General Accounting Office. "Death Penalty Sentencing: Research Indicates Pattern of Racial Disparities." GAO-GGD-90-57, February 1990.

———. "2000 Census: Coverage Measurement Programs' Results, Costs, and Lessons Learned." GAO-03-287, January 2003.

"George Gallup Jr. interview." PBS.org. http://www.pbs.org/fmc/interviews/ggallup.htm.

Gissler, Mika, et al. "Injury deaths, suicides, and homicides associated with pregnancy, Finland, 1987–2000." *European Journal of Public Health* 15, no. 5 (2005): 459–463.

Gleick, Peter H. "Hummer vs. Prius: 'Dust to Dust' Report Misleads the Media and Public with Bad Science." Pacific Institute, May 2007. http://www.pacinst.org/topics/integrity_of_science/case_studies/hummer_vs_prius.pdf.

Goldacre, Ben. "You are hereby sentenced eternally to wander the newspapers, fruitlessly mocking nutriwoo." *Guardian* (UK), July 19, 2008.

Good, I. G. "When batterer turns murderer." *Nature*, June 15, 1995, 541.

Gore v. Harris, SC00-2431 (2000).

Griffith, Elmer Cummings. *The Rise and Development of the Gerrymander.* (PhD thesis.) Chicago: Scott, Foresman and Company, 1907.

Griffith, Robert. *The Politics of Fear.* Amherst: University of Massachusetts Press, 1987.

Gross, Samuel R., et al. "Exonerations in the United States 1989 through 2003." *Journal of Criminal Law and Criminology* 95, no. 2 (2005): 523–560.

Grofman, Bernard, and Gary King. "The Future of Partisan Symmetry as a Judicial Test for Partisan Gerrymandering after *LULAC v. Perry*." *Election Law Journal* 6, no. 1 (2007): 2–35.

Hall, Mimi. "Ridge Reveals Clashes on Alerts." *USA Today*, May 10, 2005.

Halvorson, Mark, and Laura Wolff. "Report and Analysis of the 2006 Post-Election Audit of Minnesota's Voting Systems." Prepared by the Citizens for Electoral Integrity Minnesota, April 4, 2007. http://www.ceimn.org/files/CEIMNAuditReport2006.pdf.

Hamill, Pete. *Irrational Ravings*. New York: Putnam, 1971.

Hammond, William M. *Reporting Vietnam*. Lawrence: University Press of Kansas, 1998.

Harman v. Forssenius, 380 U.S. 528 (1965).

Harper v. Virginia Bd. of Elections, 383 U.S. 663 (1966).

Harsch, Joseph C. "The Politics of Peace." *Christian Science Monitor*, January 9, 1969, 18.

Hasen, Richard. "Looking for Standards (in All the Wrong Places): Partisan Gerrymandering Claims after *Vieth*." *Election Law Journal* 3, no. 4 (2004): 626–642.

Hennepin County Minnesota Voters Alliance v. City of Minneapolis, Minnesota Supreme Court A09-182 (2009).

Henry, William A., III. "Where NBC Went Wrong." *Time*, February 22, 1993.

Herrera v. Collins, 506 U.S. 390 (1993).

Hill, Ray. "Multiple sudden infant deaths—coincidence or beyond coincidence?" *Paediatric and Perinatal Epidemiology* 18 (2004): 320–326.

Hill, Russell A., and Robert A. Barton. "Red enhances human performance in contests." *Nature*, May 19, 2005, 293.

Himmelstein, David U., et al. "Illness and Injury as Contributors to Bankruptcy." *Health Affairs* Web exclusive, February 2, 2005, W5-63–W5-73. http://content.healthaffairs.org/cgi/reprint/hlthaff.w5.63v1.

Hofstadter, Richard. *Anti-Intellectualism in American Life*. New York: Vintage, 1962.

Hoppin, Jason, Dave Orrick, and Rachel Stassen-Berger. "Recount in Senate race evokes memories of 2000." *St. Paul Pioneer Press*, November 6, 2008.

Huber, Peter. "Manufacturing the Audi Scare." *Wall Street Journal*, December 18, 1989, 1.

Huff, Darrell. *How to Lie with Statistics*. New York: W. W. Norton, 1954.

"'I don't like Monday 24 January.'" BBC News, January 19, 2005. http://news.bbc.co.uk/2/hi/uk_news/4187183.stm.

"Indiana pi." Purdue University Department of Agricultural Economics Webpages. http://www.agecon.purdue.edu/crd/Localgov/Second%20Level%20pages/Indiana_Pi_Story.htm.

Jacobs v. Seminole County Canvassing Board, SC00-2447 (2000).

Jeffries, John Calvin. *Justice Lewis F. Powell Jr.* New York: Fordham University Press, 2001.

"Jesus and Princess Diana lead poll of dead people we most want to meet." Telegraph Online, June 8, 2009. http://www.telegraph.co.uk/news/newstopics/religion/5474600/Jesus-and-Princess-Diana-lead-poll-of-dead-people-we-most-want-to-meet.html.

Jones, Tim. "Dewey Defeats Truman." ChicagoTribune.com, July 13, 2008. Available at http://www.chicagotribune.com/news/politics/chi-chicagodays-deweydefeats-story,0,6484067.story.

The Journal Editorial Report. Fox News, January 10, 2009.

Just, Ward. "The Heart-Mind Gap in Vietnam War." *Washington Post*, November 19, 1967, B1.

Kaiser, Robert G. "Is This Any Way to Pick a Winner?" *Washington Post*, May 26, 2000, A1.

Kansas v. Marsh, 548 U.S. ____ (2006).

Kaszuba, Mike. "Belair still waging battle with Dziedzic." *Minneapolis Star-Tribune*, November 14, 1987, B1.

Katz, Stanley N., Hanna H. Gray, and Laurel Thatcher Ulrich. "Report of the Investigative Committee in the matter of Professor Michael Bellesiles." Emory University, July 10, 2002. Available at http://www.emory.edu/news/Releases/Final_Report.pdf.

Keating, Dan. "Lost Votes in N.M. a Cautionary Tale." *Washington Post*, August 22, 2004, A5.

Keen, Judy. "Dole still 'not certain' if tobacco is addictive; supporters are baffled by his stance." *USA Today*, July 3, 1996, A5.

"A Kerry win might be in the stars." Reuters, October 29, 2004.

"Kidnapped 'Nun Bun' resurfaces in Seattle." BoingBoing.net, March 15, 2007. http://www.boingboing.net/2007/03/15/kidnapped-nun-bun-re.html.

Knaak, Fritz. Impromptu conversation with the press in the Minnesota statehouse, November 18, 2008.

Knoll, Michelle. "Poll: Senate results unchanged if election held today," December 8, 2008. http://kstp.com/article/stories/S695337.shtml.

Kolbert, Elizabeth. "NBC Settles Truck Crash Lawsuit, Saying Test Was 'Inappropriate.'" *New York Times*, February 10, 1993, A1.

Kraushaar, Josh. "No joke: Minnesota race is sooooo ugly . . ." Politico.com, November 11, 2008. http://www.politico.com/news/stories/1108/15490.html.

"Landon, 1,293,669; Roosevelt, 972,897." *Literary Digest*, October 31, 1936, 4.

Lawrence, Felicity. "MP calls for ban on 'unsafe' sweetener." *Guardian* (UK), December 15, 2005.

League of United Latin American Citizens v. Perry, 548 U.S. 399 (2006).

Leake, Jonathan. "Physicists fear Big Bang machine could destroy Earth." *Calgary Herald*, July 19, 2000, A7.

Lee, Margaret Mikyung. "Sampling for Census 2000: A Legal Overview." Library of Congress, Congressional Research Service, Report CRS-97-871, updated December 30, 1998.

Lee's Summit v. Surface Transportation Board, 231 F.3d 39; 343 U.S. App. D.C. 405 (2000).

Lester, Will. "'Dewey Defeats Truman' Disaster Haunts Pollsters." Associated Press, November 1, 1998.

———. "Poll: Abandon Flooded New Orleans." Associated Press, September 9, 2005.

Lewis, Anthony. "Bowing to Racism." *New York Times*, April 28, 1987, A31.

Lewis, Michael. "The Man Who Crashed the World." *Vanity Fair*, August 2009.

Limbaugh, Rush. *The Rush Limbaugh Show*, March 13, 2007.

Lindgren, James. "Fall From Grace: *Arming America* and the Bellesiles Scandal." *Yale Law Journal*, June 2002, 2195–2249.

Lippmann, Walter. *Public Opinion*. Part VII, chapter XXIII.

Lopez, Patricia. "Franken's Deficit: 236 Votes." *Minneapolis Star-Tribune*, November 7, 2008.

LULAC v. Perry, 548 U.S. ____ (2006).

Mackowiak, Philip A., et al. "A Critical Appraisal of 98.6°F, the Upper Limit of the Normal Body Temperature, and Other Legacies of Carl Reinhold August Wunderlich." *Journal of the American Medical Association*, September 23/30, 1992, 1578–1580.

Madsen, K. M., et al. "Thimerosal and the occurrence of autism: negative ecological evidence from Danish population-based data." *Pediatrics*, September 2003, 604–606.

Magazine advertisement for Heinz Tomato Juice. *Literary Digest*, November 14, 1936, 3.

"Mann & Woman." *Time*, April 3, 1944.

Manning, J. T., et al. "The mystery of female beauty." *Nature*, May 20, 1999, 214–215.

———. "Women's waists and the sex ratio of their progeny: evolutionary aspects of the ideal body shape." *Journal of Human Evolution* 31 (1996): 41–47.

"Marijuana and your teen's mental health." Magazine advertisement, Office of National Drug Control Policy, ca. 2005.

Marquis, Joshua. "The Innocent and the Shammed." *New York Times*, January 26, 2006, A23.

———. "The Myth of Innocence." *Journal of Criminal Law and Criminology* 95, no. 2 (2005): 501–522.

Martis, Kenneth C. "The Original Gerrymander." *Political Geography*, November 2008, 833–839.

"M'Carthy Insists Truman Oust Reds." *New York Times*, February 12, 1950, 5.

McCleskey v. Kemp, 481 U.S. 279 (1987).

Meadow, Roy. "A case of murder and the *BMJ*." *British Medical Journal*, January 5, 2002, 41–43.

———. "Unnatural Sudden Infant Death." *Archives of Disease in Childhood* 80 (1999): 7–14.

Medina, Jennifer. "Little Progress for City Schools on National Test." *New York Times*, November 16, 2007.

———. "Reading and Math Scores Rise Sharply Across N.Y." *New York Times*, June 24, 2008.

Miller, Judith. "Defectors Bolster U.S. Case Against Iraq, Officials Say." *New York Times*, January 24, 2003, A11.

———. "Experts Split on Flaws in Pacts Limiting Nuclear Tests." *New York Times*, July 26, 1982, A3.

———. "U.S. Confirms a Plan to Halt Talks on a Nuclear Test Ban." *New York Times*, July 21, 1982, A1.

———. "U.S. Said to Decide Against New Talks to Ban All A-Tests." *New York Times*, July 20, 1982, A1.

Minnesota Election Laws 2008.

Minnesota State Canvassing Board meeting, November 18, 2008.

Minnesota State Canvassing Board meeting, December 18, 2008.

"Mischief in Minnesota." *Wall Street Journal*, November 12, 2008.

Mooney, Chris. "John Zogby's Creative Polls." *American Prospect*, February 1, 2003.

NAACP v. Harris, 01-120-CIV-GOLD/SIMONTON (2002).

NAEP Data Explorer. http://nces.ed.gov/nationsreportcard/naepdata/.

NAMUDNO v. Holder, 557 U.S. ____ (2009).

"National Security Agency." RasmussenReports.com, survey of 1000 adults, December 26–27, 2005. http://legacy.rasmussenreports.com/2005/NSA.htm.

"New AIG CEO Discusses Company Spending, Troubles." *NewsHour with Jim Lehrer*, October 22, 2008. Transcript available at http://www.pbs.org/newshour/bb/business/july-dec08/aigcrisis_10-22.html.

"The New York Times/CBS News Poll." January 20–25, 2006. http://www.nytimes.com/packages/pdf/politics/20060127_poll_results.pdf.

Noah, Timothy. "The Bellesiles of the Right?" Slate.com. http://www.slate.com/toolbar.aspx?action=print&id=2078084.

"The official website of the Beijing 2008 Olympic Games." http://results.beijing2008.cn/WRM/ENG/Schedule/. (currently defunct).

Olney, J. W., et al. "Increasing Brain Tumor Rates: Is There a Link to Aspartame?" *Journal of Neuropathology and Experimental Neurology*, November 1996, 1115–1123.

O'Loughlin, John. "The Identification and Evaluation of Racial Gerrymandering." *Annals of the Association of American Geographers*, June 1982, 165–184.

Oshinsky, David M. *A Conspiracy So Immense*. New York: Free Press, 1983.

Palm Beach County Canvassing Board v. Harris, SC00-2346 (2000).

Paulos, John Allen. *A Mathematician Reads the Newspaper*. New York: Basic Books, 1995.

———. *Innumeracy*. New York: Hill and Wang, 1988.

———. "Murder he wrote." http://www.math.temple.edu/~paulos/oj.html.

Paulsen, Michael Stokes. "The Minnesota Recount Was Unconstitutional." *Wall Street Journal*, January 15, 2009.

Peck, Don, and Caitlin Casey. "Packing, Cracking, and Kidnapping." *Atlantic Monthly*, January/February 2004, 50–51.

People v. Collins, 68 Cal. 2d 319; 438 P.2d 33; 66 Cal. Rptr. 497 (1968).

"Plans for spaceport Sweden." Press release, virgingalactic.com, April 3, 2008. http://www.virgingalactic.com/htmlsite/news.php.

Presidential Commission on the Space Shuttle Accident. *Report of the Presidential Commission on the Space Shuttle Accident*, June 6, 1986.

Pyle, Richard, and Horst Faas. *Lost over Laos*. Cambridge, MA: Da Capo, 2003.

Ragsdale, Jim. "A Statistical Dead Heat, Heating Up." *St. Paul Pioneer Press*, November 12, 2008.

Reed, Henry, and Charles Thomas Cayce. *Awakening Your Psychic Powers*. New York: Macmillan, 1988.

"Researchers Link Bad Debt to Bad Health." *New York Times*, March 14, 2000, D8.

Reynolds v. Sims, 377 U.S. 533 (1964).

Risinger, D. Michael. "Innocents Convicted: An Empirically Justified Factual Wrongful Conviction Rate." *Journal of Criminal Law and Criminology* 97, no. 3 (2007): 761–805.

Roberts, L. Clayton. Memo to Al Cardenas, "Definitions of Errors in Vote Tabulation," DE-00-11, November 13, 2000. http://election.dos.state.fl.us/opinions/new/2000/de0011.pdf.

———. Memo to Charles E. Burton, "Deadline for Certification on County Results," DE-00-10, November 13, 2000. http://election.dos.state.fl.us/opinions/new/2000/de0010.pdf.

Rookus, Matti A., and Flora E. van Leeuwen. "Induced Abortion and Risk for Breast Cancer: Reporting (Recall) Bias in a Dutch Case-Control Study." *Journal of the National Cancer Institute*, December 4, 1996, 1759–1764.

"Rose Parade: Big crowd, but who's counting?" LATimes.com, January 6, 2009. http://latimesblogs.latimes.com/readers/2009/01/rose-parade-big.html.

Rucker, Philip. "Activists bring 'tea party' to Capitol Hill." *Washington Post*, November 6, 2009.

Saad, Lydia. "Late upsets are rare, but have happened." Gallup.com. http://www.gallup.com/poll/111451/late-upsets-rare-happened.aspx.

Safire, William. Memo to H. R. Haldeman, "In Event of Moon Disaster," July 18, 1969. http://www.thesmokinggun.com/archive/apollo1.html.

Sagan, Scott D. "Nuclear Alerts and Crisis Management." *International Security*, Spring 1985, 99–139.

Sayer, Peter. "Study: Electronic Voting Increased Counting Errors in France." PCWorld.com, July 9, 2008. http://www.pcworld.com/printable/article/id,148112/printable.html.

Schuon, Marshall. "Audi Redesigns and Presses On." *New York Times*, April 18, 1993, section 8, p. 14.

Schwartz, John. "Report Linking Sweetener to Brain Cancer is Disputed; FDA Finds No Reason to Question Aspartame's Safety." *Washington Post*, November 19, 1996, A2.

"Security Risks Kept in Jobs, Senator Says." *Washington Post*, February 12, 1950, M1.

Seife, Charles. "Columbia Disaster Underscores the Risky Nature of Risk Analysis." *Science*, February 14, 2003, 1001–1002.

———. "Gambling with Our Votes." *Science*, October 29, 2004, 798–799.

———. "Increasing Brain Tumor Rates: Is There a Link to Deficit Spending?" *Journal of Neuropathology and Experimental Neurology*, April 1999, 404–405.

———. "Not Every Vote Counts." *New York Times*, December 4, 2008, A43.

———. "Pie in the sky." *New Scientist*, April 10, 1999, 1414.
———. "Predicting the Presidents." *Washington Post*, June 3, 2000, A15.
Shapiro, Jacob N., and Dara Kay Cohen. "Color Bind: Lessons from the Failed Homeland Security Advisory System." *International Security*, Fall 2007, 121–154.
Shenon, Philip, and Eric Lichtblau. "U.S. Lowers Warning Level to 'Yellow,' but Cautions That Serious Threat Remains." *New York Times*, February 28, 2003, A14.
"She's one in 300,000,000." *Chicago Sun-Times*, October 18, 2006, 6.
Silk, Leonard. "Forecasting tool gives nod to Bush." *New York Times*, March 20, 1992, D2.
Silver, Nate. "Projection: Franken to win by 27 votes." FiveThirtyEight.com, November 23, 2008. http://www.fivethirtyeight.com/2008/11/projection-franken-to-win-recount-by-27.html.
———. "Statistical Models Now Show Coleman as Slight Favorite." FiveThirtyEight.com, December 2, 2008. http://www.fivethirtyeight.com/2008/12/statistical-models-now-show-coleman-as.html.
Singh, Devandra, and Robert J. Zambrano. "Offspring sex ratio in women with android body fat distribution. *Human Biology* 69 (1997): 545–556.
"Sir Roy Meadow struck off by GMC." BBC News, July 5, 2005. http://news.bbc.co.uk/go/pr/fr/-/1/hi/health/4685511.stm.
Smith, Leef, and Wendy Melillo. "If It's Crowd Size You Want, Park Service Says Count It Out." *Washington Post*, October 13, 1996, A34.
South Dakota Codified Laws.
"Spaceport America Breaks Ground in New Mexico." Press release, spaceportamerica.com, June 22, 2009. http://www.spaceportamerica.com/news/press-releases/244-spabreaksgroundnewmexico.html
Squire, Peverill. "Why the 1936 *Literary Digest* Poll Failed." *Public Opinion Quarterly*, Spring 1988, 125–133.
Stassen-Berger, Rachel, and Jason Hoppin. "In Senate recount, wrongly rejected votes seem more numerous than first thought." *St. Paul Pioneer Press*, December 11, 2008.
Steinberg, Earl P. "The Status of MRI in 1986: Rates of Adoption in the United States and Worldwide." *American Journal of Radiology*, September 1986, 453–455.
Sternberg, William. "Cooked Books." *Atlantic Monthly*, January 1992, 20.
Stewart v. Blackwell, No. 05-3044, Sixth Circuit Court of Appeals (2006).
Stille, Alexander. "How the Press Saw the 2000 Election." *New York Times*, January 8, 2003.
Sunstein, Cass, and Justin Wolfers. "A Death Penalty Puzzle." *Washington Post*, June 30, 2008, A11.
Sunstein, Cass R., and Adrian Vermeule. "Is Capital Punishment Morally Required? Acts, Omissions, and Life-Tradeoffs." *Stanford Law Review* 58, no. 3 (2005): 703–750.
———. "Deterring Murder: A Reply." *Stanford Law Review* 58, no. 3 (2005): 847–857.
Superville, Darlene. "Poll: Americans see doom, gloom for 2007." Associated Press, December 31, 2006.
Sweetland, Anders. *Item Analysis of the HES (Hamlet Evaluation System)*. 06-M-0004. Santa Monica, CA: RAND Corporation, August 20, 1968.
Tapper, Jake. "Buchanan camp: Bush claims are 'nonsense.'" Salon.com, November 10, 2000. http://dir.salon.com/story/politics/feature/2000/11/10/buchanan/.

Tatem, Andrew J., et al. "Momentous sprint at the 2156 Olympics?" *Nature*, September 30, 2004, 525.

Taylor, Andrew. "AIG execs' retreat after bailout angers lawmakers." Associated Press, October 7, 2008.

Taylor v. Martin County Canvassing Board, SC00-2448 (2000).

Television advertisement for L'Oreal Extra Volume Collagen Mascara, http://www.lorealparisusa.com/.

Television advertisement for Vaseline Dry Skin, http://www.vaseline.com/.

Texas Code of Criminal Procedure.

Texas Defender Service. "A State of Denial: Texas Justice and the Death Penalty," October 16, 2000.

Texas Department of Criminal Justice. "Last Statement—Thomas Barefoot." http://www.tdcj.state.tx.us/stat/barefootthomaslast.htm.

Texas Health and Safety Code.

Thayer, Thomas C., ed. *A Systems Analysis View of the Vietnam War 1965–1972*, vols. 9 and 10. OASD(SA)RP Southeast Asia Intelligence Division, Pentagon, Washington, D.C., February 18, 1975.

"Thefts puncture Paris bike scheme." BBC News, February 10, 2009. http://news.bbc.co.uk/2/hi/europe/7881079.stm.

"They Call Him Dr. Death." *Time*, June 1, 1981.

Thompson, William C., and Edward L. Schumann." Interpretation of Statistical Evidence in Criminal Trials: The Prosecutor's Fallacy and the Defense Attorney's Fallacy." *Law and Human Behavior*, September 1987, 167–187.

"The Times and Iraq." *New York Times*, May 26, 2004. Available at http://www.nytimes.com/2004/05/26/international/middleeast/26FTE_NOTE.html?pagewanted=all.

"Times Poll: Frequently Asked Questions." LATimes.com. http://www.latimes.com/news/custom/timespoll/la-timespollfaq,1,2370162.htmlstory.

Toobin, Jeffrey. "The Great Election Grab." *New Yorker*, December 8, 2003.

Tovee, M. J., J. E. Brown, and D. Jacobs. "Maternal waist-to-hip ratio does not predict child gender." *Proceedings of the Royal Society of London B* 268 (2001): 1007–1010.

Torricelli, Robert G., and Andrew Carroll. *In Our Own Words*. New York: Washington Square Press, 1999.

Tuck, Angela. "'Objective truth' passes the test for letters to the editor." *Atlanta Journal-Constitution*, August 4, 2007, A15.

Tversky, Amos, and Daniel Kahneman. "The Framing of Decisions and the Psychology of Choice." *Science*, January 30, 1981, 453–458.

"'Twinkiegate' Candidate Faces Charges." Associated Press, May 23, 1986.

"Twinkiegate Case Closed." *Houston Chronicle*, June 5, 1986, 11.

"2008 Beijing Olympic Games Website." http://www.2008.nbcolympics.com/.

United States of America v. Russell Cletus Maricle et al., No. 09-16-S-DCR. U.S. District Court, Eastern District of Kentucky, Southern Division, London (2009).

U.S. Commission on Civil Rights. "Voting Rights Enforcement & Reauthorization," May 2006.

U.S. Congress. House of Representatives, Committee on Oversight and Government

Reform. "Hearings on the Causes and Effects of the AIG Bailout," October 7, 2008.

U.S. Department of State Bulletin, February 1986. Available at http://findarticles.com/p/articles/mi_m1079/is_v86/ai_4115455/.

U.S. House of Representatives v. U.S. Department of Commerce et al., Civ. A. No. 98-0456.

Utah v. Evans, 536 U.S. 452 (2002).

van der Vink, Gregory E. "The Role of Seismologists in Debates over the Comprehensive Test Ban Treaty." *Annals of the New York Academy of Sciences* 866 (1998): 84–113.

Venkataraman, Nitya. "Bush in Retrograde; Obama, McCain Rising." ABC News, June 2, 2008. http://abcnews.go.com/Politics/Vote2008/story?id=5286338&page=1.

Vieth v. Jubelirer, 541 U.S. 267 (2004).

von Sternberg, Bob. "A pre-count for the Coleman-Franken recount." *Minneapolis Star-Tribune*, November 10, 2008.

Voting Rights Act of 1965.

Wallace, Nicole. "Donations to Hurricane-Relief Efforts Exceed $504-Million." *Chronicle of Philanthropy*, September 6, 2005.

Waxman, Henry A., et al. Letter to Tom Ridge, June 4, 2003. http://oversight.house.gov/images/stories/documents/20040628114808-24103.pdf.

Weiner, Jay. "A blustery Recount Saturday: 'We win' by 35–50 votes, says Franken lawyer; 'Hot air,' replies Coleman spokesman." MinnPost.com, December 20, 2008. http://www.minnpost.com/stories/2008/12/20/5443/a_blustery_recount_s.

Weir, Bruce S. "DNA statistics in the Simpson matter." *Nature Genetics*, December 11, 1995, 365–368.

"We're number one, alas." *Wall Street Journal*, July 13, 2007, A12.

Whipp, Brian J., and Susan A. Ward. "Will women soon outrun men?" *Nature*, January 2, 1992, 25.

Whitman, Mark, ed. *Florida 2000: A Sourcebook on the Contested Presidential Election*. Boulder, CO: Lynne Rienner Publishers, 2003.

"Why will Virgin Galactic offer a safer way to travel to space (and back!)." virgingalactic.com. http://www.virgingalactic.com/htmlsite/faq.php?subtitle=Safety&src=141.

Will, George F. "Fuzzy Climate Math." *Washington Post*, April 12, 2007, A27.

Winerip, Michael. "One Secret to Better Test Scores: Make State Reading Tests Easier." *New York Times*, October 5, 2005.

"Zogby America Likely Voters 3/30/05 thru 4/2/05 MOE +/– 3.2 percentage points." Zogby.com, April 2, 2005. http://operationrescue.org/files/wf-Schiavo.pdf.

"Zogby Poll: Americans Not in Favor of Starving Terri Schiavo," Zogby.com, April 1, 2005. http://www.zogby.com/Soundbites/readclips.cfm?ID=11131.

Index